AF360703

Proteases—From Basic Structure to Function to Drug Design as Targeted Therapy

Proteases—From Basic Structure to Function to Drug Design as Targeted Therapy

Editors

Hang Fai Kwok
Christopher Shaw
Brian Walker

MDPI • Basel • Beijing • Wuhan • Barcelona • Belgrade • Manchester • Tokyo • Cluj • Tianjin

Editors

Hang Fai Kwok
Department of Biomedical
Sciences
Faculty of Health Sciences
University of Macau
Taipa
Macau

Christopher Shaw
School of Pharmacy
Queen's University Belfast
Belfast
United Kingdom

Brian Walker
School of Pharmacy
Queen's University Belfast
Belfast
United Kingdom

Editorial Office
MDPI
St. Alban-Anlage 66
4052 Basel, Switzerland

This is a reprint of articles from the Special Issue published online in the open access journal *Biology* (ISSN 2079-7737) (available at: www.mdpi.com/journal/biology/special_issues/proteases).

For citation purposes, cite each article independently as indicated on the article page online and as indicated below:

LastName, A.A.; LastName, B.B.; LastName, C.C. Article Title. *Journal Name* **Year**, *Volume Number*, Page Range.

ISBN 978-3-0365-2575-4 (Hbk)
ISBN 978-3-0365-2574-7 (PDF)

Contents

Preface to "Proteases—From Basic Structure to Function to Drug Design as Targeted Therapy" vii

Anastasia S. Frolova, Anastasiia I. Petushkova, Vladimir A. Makarov, Surinder M. Soond
and Andrey A. Zamyatnin
Unravelling the Network of Nuclear Matrix Metalloproteinases for Targeted Drug Design
Reprinted from: *Biology* 2020, *9*, 480, doi:10.3390/biology9120480 1

Etienne J. Slapak, JanWillem Duitman, Cansu Tekin, Maarten F. Bijlsma and C. Arnold Spek
Matrix Metalloproteases in Pancreatic Ductal Adenocarcinoma: Key Drivers of Disease
Progression?
Reprinted from: *Biology* 2020, *9*, 80, doi:10.3390/biology9040080 21

Jaana Künnapuu, Honey Bokharaie and Michael Jeltsch
Proteolytic Cleavages in the VEGF Family: Generating Diversity among Angiogenic VEGFs,
Essential for the Activation of Lymphangiogenic VEGFs
Reprinted from: *Biology* 2021, *10*, 167, doi:10.3390/biology10020167 43

Pavla Bartošová-Sojková, Jiří Kyslík, Gema Alama-Bermejo, Ashlie Hartigan, Stephen D.
Atkinson, Jerri L. Bartholomew, Amparo Picard-Sánchez, Oswaldo Palenzuela, Marc Nicolas
Faber, Jason W. Holland and Astrid S. Holzer
Evolutionary Analysis of Cystatins of Early-Emerging Metazoans Reveals a Novel Subtype in
Parasitic Cnidarians
Reprinted from: *Biology* 2021, *10*, 110, doi:10.3390/biology10020110 67

Tao Wang, Yangyang Jiang, Xiaoling Chen, Lei Wang, Chengbang Ma, Xinping Xi, Yingqi
Zhang, Tianbao Chen, Chris Shaw and Mei Zhou
Ranacyclin-NF, a Novel Bowman–Birk Type Protease Inhibitor from the Skin Secretion of the
East Asian Frog, *Pelophylax nigromaculatus*
Reprinted from: *Biology* 2020, *9*, 149, doi:10.3390/biology9070149 83

Preface to "Proteases—From Basic Structure to Function to Drug Design as Targeted Therapy"

In the two last decades, proteases have constituted one of the primary and important targets in drug discovery. The U.S. FDA has approved more than 12 protease therapies in the last 10 years, and a number of next-generation or completely new proteases are under clinical development. Protease inhibition strategies are one of the fastest expanding areas in the field of drugs that show considerable promise. This Special Issue will focus on the recent advances in the discovery and development of protease inhibitors, covering the synthesis of protease inhibitors, the design of new chemical entities acting as inhibitors of special/particular types of proteases, and their mode of actions (Frolova et al. 2020; Slapak et al. 2020; Künnapuu et al. 2021). In addition, the new applications of these interesting compounds/biomolecules and their limitations have been discussed and described (Wang et al. 2020; Bartošová-Sojková et al, 2021).

The editors are grateful to all the authors who contributed to this Special Issue *"Proteases—From Basic Structure to Function to Drug Design as Targeted Therapy"*. They are also mindful that without the rigorous and selfless evaluation of the submitted manuscripts by external peer reviewers/expertise, this Special Issue could not have happened. Moreover, the editor (H. F. KWOK) gives thanks for the support from the Science and Technology Development Fund of Macau SAR (FDCT) (file no. 0055/2019/A1 and 0010/2021/AFJ) and the Faculty of Health Sciences (FHS) University of Macau. Finally, the valuable contributions, organization, and editorial support of the MDPI management team and staff are greatly appreciated.

Hang Fai Kwok, Christopher Shaw, Brian Walker
Editors

Review

Unravelling the Network of Nuclear Matrix Metalloproteinases for Targeted Drug Design

Anastasia S. Frolova [1], **Anastasiia I. Petushkova** [1], **Vladimir A. Makarov** [1], **Surinder M. Soond** [1] and **Andrey A. Zamyatnin Jr.** [1,2,3,*]

[1] Institute of Molecular Medicine, Sechenov First Moscow State Medical University, 119991 Moscow, Russia; frolanasta@gmail.com (A.S.F.); asyapeti@gmail.com (A.I.P.); known.sir@yandex.ru (V.A.M.); surinder.soond@yandex.ru (S.M.S.)

[2] Belozersky Institute of Physico-Chemical Biology, Lomonosov Moscow State University, 119992 Moscow, Russia

[3] Department of Biotechnology, Sirius University of Science and Technology, 1 Olympic Ave., 354340 Sochi, Russia

* Correspondence: zamyat@belozersky.msu.ru; Tel.: +7-495-622-98-43

Received: 7 November 2020; Accepted: 17 December 2020; Published: 19 December 2020

Simple Summary: Nuclear matrix metalloproteinases are emerging to have distinct functions in a number of pathological conditions and physiological processes. In this article, we review what progress has been made in this area of research and discuss their potential in being targeted for future therapeutic design.

Abstract: Matrix metalloproteinases (MMPs) are zinc-dependent endopeptidases that are responsible for the degradation of a wide range of extracellular matrix proteins, which are involved in many cellular processes to ensure the normal development of tissues and organs. Overexpression of MMPs has been observed to facilitate cellular growth, migration, and metastasis of tumor cells during cancer progression. A growing number of these proteins are being found to exist in the nuclei of both healthy and tumor cells, thus highlighting their localization as having a genuine purpose in cellular homeostasis. The mechanism underlying nuclear transport and the effects of MMP nuclear translocation have not yet been fully elucidated. To date, nuclear MMPs appear to have a unique impact on cellular apoptosis and gene regulation, which can have effects on immune response and tumor progression, and thus present themselves as potential therapeutic targets in certain types of cancer or disease. Herein, we highlight and evaluate what progress has been made in this area of research, which clearly has some value as a specific and unique way of targeting the activity of nuclear matrix metalloproteinases within various cell types.

Keywords: matrix metalloproteinase; extracellular matrix; nuclei; cancer; apoptosis; immune response

---

1. Introduction

Matrix metalloproteinases (MMPs) are involved in the degradation of extracellular matrix (ECM) proteins and regulate many fundamental cellular processes during normal bodily development and function [1]. As the ECM is important in maintaining the mechanical and biochemical properties of tissues, its normal turnover and regulation by MMPs is necessary to permit multiple functions, as in the cleavage and activation of signaling molecules, cellular differentiation, and wound healing [2–7]. However, dysregulation of MMP activity can contribute to a variety of pathological conditions. For example, some have been seen to modulate matrix erosion in osteoarthritis and rheumatoid arthritis, whereas expression of others is associated with the formation of atherosclerotic lesions, platelet aggregation, and the regulation of factors associated with cardiovascular disease [8,9]. Predominantly, the roles of MMPs in malignant tumor initiation, metastasis, and angiogenesis have received the greatest attention and which have highlighted them as good potential therapeutic targets for the treatment of certain types of cancer [9].

To date, 26 human MMP proteins have been identified, which belong to the M10 family of metallo-endopeptidases [10]. Based on substrate specificity, MMPs can be further categorized into collagenases (MMP-1, MMP-8, MMP-13, and MMP-18), gelatinases (MMP-2 and MMP-9), stromelysins (MMP-3, MMP-10, MMP-11, and MMP-17), matrilysins (MMP-7 and MMP-26), membrane-type MMPs (MMP-14, MMP-15, MMP-16, MMP-17, MMP-24, and MMP-25), and others (MMP-12, MMP-19, MMP-20, MMP-21, MMP-22, MMP-23, MMP-28, and MMP-29) [1]. Generally speaking, they are expressed by a broad range of cell types, such as epithelial cells, fibroblasts, osteoblasts, endothelial cells, vascular smooth muscle, macrophages, neutrophils, lymphocytes, and cytotrophoblasts [1].

Structurally, MMPs share a common protein domain structure (Figure 1). For most MMPs, the main components are a signal peptide (that directs synthesized protein into the secretory pathway), a highly conserved amino-terminal pro-domain, a catalytic domain that contains a zinc ion binding site, a linker domain, and a carboxyl-terminal hemopexin-like domain (HEX), that determines substrate specificity and localization and contributes to the enzymatic activity of MMPs [11].

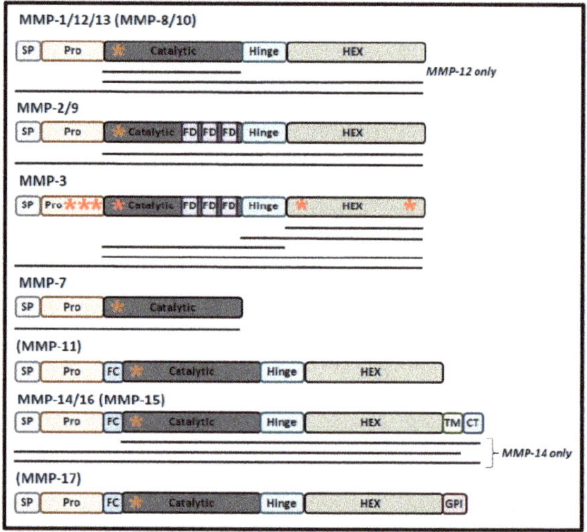

Figure 1. Domain structures of human matrix metalloproteinases (MMPs). Here the domain structures of nuclear MMPs (nMMPs) and MMPs, which have not been found in the nuclei but possess nuclear localization signals (NLSs) (placed in brackets), are presented. NLSs are indicated by the red asterisks (if their nuclear-trafficking properties have been proven experimentally) or by the orange asterisks (if the NLS has been identified by bioinformatics alone). Horizontal lines indicate the isoforms of MMPs, which have been found in nuclei. SP, signal peptide; Pro, pro-domain; FC, furin cleavage site; FD, fibronectin domain; HEX, hemopexin-like domain; TM, trans-membrane domain; CT, cytoplasmic tail; GPI, glycosylphosphatidylinositol.

These proteases are synthesized in the form of pre-pro-MMPs, with their enzymic activation occurring through the process of maturation as the proteins progress through the secretory pathway [1]. The first step of maturation is removal of the secretory signal peptide following the course of protein translation, giving rise to an inactive pro-MMP in which the inhibition of the catalytic site occurs through its resident Zn^{2+} ion binding a cysteine residue within the "cysteine switch" motif (PRCGXPD) present in the pro-domain [12]. Activation of the pro-MMP may occur in a variety of different ways, arising in a number of MMP forms containing the full-length pro-domain, a processed form of the pro-domain, and in MMPs lacking the pro-domain. In the former two MMP derivatives, conformational changes caused by mechanical or chaotropic agents can lead to the disruption of the Zn^{2+}-Cys interaction resulting in pro-MMP activation in the absence of pro-domain cleavage [13]. Moreover, processed cleavage and removal of the pro-domain, by plasmin or trypsin, can mediate a conformation change of the protease resulting in full activation of the MMP intermediate [13]. Normally, full cleavage of the pro-domain is either mediated by the furin pro-protein convertase in the trans-Golgi network, auto-catalytically, or by other MMPs at the cell's surface, either within the ECM or the nucleus [12,14,15]. The activity of MMPs can also be regulated by post-translational modifications, such as glycosylation, phosphorylation, and by glycosaminoglycans (GAGs). For example, glycosylation can stabilize a complex between MMP-14, TIMP2, and pro-MMP-2 as a step necessary for the cell-surface activation of MMP-2 [16]. Alternatively, glycosylation can promote MMP-9 secretion and activation, while also stabilizing the formation of MMP-17 dimers [17]. As an important step for the classical mode of MMP activation, a number of recent studies have also reported that some MMPs are also responsive to redox-mediated activation [18].

The tissue inhibitors of metalloproteinase (TIMPs) have also gained significant importance over the years based on their developmental role in normal tissue homeostasis and disease progression and their abilities to modulate MMP protease activity [19]. Four TIMPs (TIMPs 1-4) have been identified, and their mechanisms of MMP inhibition have been established through a number of structural studies. Residues 1–4 of the TIMP-1 amino-terminal domain interact with the primed side of the MMP binding pocket, where Cys-1 can coordinately bind the catalytic site Zn^{2+} ion. Simultaneously, five residues (spanning amino acids 66–70) from TIMP-1 can occupy the non-primed site [20]. These potential modes of binding were also shown to be highly conserved among TIMP-2, TIMP-3, and TIMP-4 [20,21]. Biologically, elevated TIMP expression levels have been shown to contribute to enhanced ECM accumulation and deposition, while reduced TIMP expression leads to enhanced matrix proteolysis, thus highlighting their importance in modulating ECM dynamics and plasticity [1,22]. TIMPs can also form non-inhibitory pro-MMP/TIMP/MT-MMP complexes, as in the instance of TIMP-2 complexing with MMP-14 and which can activate pro-MMP-2 in human fibrosarcoma, breast, and melanoma cell lines [16]. While TIMPs are generally found within the ECM, a number of studies have demonstrated that they may also reside in the nucleus of cells, as seen for TIMP-1 [23–25].

Over the years, matrix metalloproteinases have been pursued as good targets for therapeutic development [9], and have the potential to be targeted at several levels of their synthesis and maturation, the proposed stages of which include inhibition at the transcriptional level, during zymogen activation, and at the level of substrate catalysis by the active enzyme [9]. At the moment, there are MMP-directed targeted strategies coming into fruition for the treatment of inflammation, heart disease, lung diseases, and ischemic stroke [26–30]. Simultaneously, the search for more specific and better MMP inhibitors is still ongoing, driven by limited options for targeting specific MMPs within a clinical setting [31]. Consequently, novel strategies embodying greater specificity and efficacy have taken on a greater priority in targeting MMPs.

Over the last ten years, the nuclear localization of MMPs (nMMPs) has been an increasingly reported phenomenon, which has been observed in high-grade tumors, correlated with tumor volume, and in some instances has been associated with poor prognosis in a number of disease types (Table 1) [32–36]. Collectively, such findings suggest an important functional role for nuclear MMPs and that such a localization effect does have biological and clinical significance. In support of this, it is interesting to note that nuclear localization has been reported for other ECM proteases as well. For example, nuclear cathepsins L and D have been reported to exhibit biological effects which can contribute to tumor progression [37–39]. Collectively, the localization of such proteases have the potential to activate or deactivate transcription factors, regulate chromatin remodeling, apoptosis, alter the structural elements of the nuclear matrix, and participate in molecular events that lead to cell proliferation and carcinogenesis [40–43].

What signaling cues cause MMPs to be directed to the nucleus still largely remains unknown, with a number of mechanisms being proposed, which stem from environmental factors to cellular metabolism [44]. Nevertheless, for several MMPs, researchers have been able to propose some molecular mechanisms responsible for nuclear MMP translocation [45–52].

In this review article, we highlight the increasing emergence of nMMPs, while outlining their biological significance to highlight how these distinct sub-sets of proteases may have good targeting potential in diseases such as cancer.

2. Mechanistic Regulation of Nuclear MMPs

One of the commonest ways to deliver proteins from the cytoplasm to the nucleus is through the process of receptor-mediated nuclear shuttling and due (in large part) to proteins possessing a nuclear localization sequence (NLS) [53]. Here, importins α and β recognize and bind the NLS to form an importin-cargo complex, which can bind the nuclear pore complex, and facilitate the translocation of protein cargo from the cytoplasm into the nucleus [54]. Depending on its sequence and structure, the NLS can be sub-divided into two groups composed of the classical NLS and the proline-tyrosine (PY) NLS. Specific importin proteins are also involved in recognizing different types of NLS, which allows them to confer protein selectivity and regulate this mechanism with greater specificity [53].

The classical NLS was originally thought to be involved in the nuclear translocation of MMP-2 when this protease was detected in the nucleus of rat cardiac myocyte cells for the first time [55]. This was confirmed upon scrutinizing the rat MMP-2 protein sequence, which revealed two small stretches of basic amino acids close to the C-terminal separated by a variable spacer [55]. Such sequences were also identified in the catalytic domain of MMP-3 [56]. The putative NLS (PKWRKTH) was identified and confirmed using the bioinformatics software, Protein Subcellular Localization Prediction Tool (PSORT, https://psort.hgc.jp/) [57], and validated upon the deletion of two positively-charged amino acids from this putative NLS, which led to a large decrease in the nuclear localization of the mutated proteins [56]. The same outcomes were observed after the substitution of these amino acids with uncharged amino acids as in the amino acid substitutions **R**110N and **K**111Q. For the first time in this field of research, such findings demonstrated a potential molecular mechanism for the nuclear translocation of MMP-3 [56]. Subsequently, Eguchi et al. (2008) identified five additional putative NLSs in MMP-3, of lysine- and arginine-rich sequences and which were found to be dispersed throughout all of the MMP-3 protein domains (Figure 2) [58]. Moreover, all of these NLSs were exclusively able to transport the MMP-3 protein into the nucleus. While such a study highlighted the existence of multiple NLSs and their dispensability, it also suggested that the post-translational modification of MMPs may "hide" primary NLSs in addition to exposing alternative NLSs and which may offer potential mechanisms that confer selectivity for the nuclear shuttling of some proteins. Functionally, nMMP-3 has also been shown to participate in the transcriptional regulation of *CTGF/CCN2* and *HSP* gene expression, where the presence (or absence) of each of the NLSs may contribute to regulating TG2, ERK, and IL-33 specific signaling pathways and responses [58–60].

The use of bioinformatic analyses have also helped to develop this area of research through identifying additional putative NLS sequences in other human MMPs protein sequences. Here, Abdukhakimova et al. (2016) identified a putative NLS within the catalytic domain of 14 MMPs, including the above-mentioned MMP-2 and MMP-3 proteins [61]. The sequences of MMPs were compared with experimentally validated NLSs from the catalytic domain of MMP-3 (of sequence PKWRKTH) [58] and most of the recovered NLSs contained two consensus residues, namely lysine and tryptophan (KW). The whole sequence was identified only in MMP-3 and MMP-10 and the authors also revealed the importance of the NLS in MMP-7 through it being evolutionary conserved throughout different species (Figure 2) [50,61].

Mechanistically, it has been proposed that endocytosis may also be responsible for the nuclear localization of MMPs. For example, in hepatocellular carcinoma cells, the amount of nuclear MMP-14 protein was increased in comparison to healthy liver cells, an event which enhanced the metastatic capacity of tumor cells [33]. Here, MMP-14 was jointly localized within the cytoplasm and perinuclear space and could interact with caveolin-1, thus implicating a specialized form of endocytic-protein trafficking that is fundamentally different to the use of nuclear transport receptors [54,62]. In support of this, caveolin-1 has been reported to drive and enrich the transport of proteins to the nucleus in human endothelial cells, as seen with caveolae regulating the intracellular protein trafficking of MMP-14 [63,64]. Such observations enforce the proposition that caveolin-1 participates in the nuclear translocation of MMP-14.

Classical NLS	K(K/R)X(K/R)
MMP-3 pro1	44-LKKDVKQFVRRKD-56
MMP-3 pro2	59-PVVKKIR-65
MMP-3 pro3	84-LEVMRKPR-91
MMP-3 hex1	312-ILIFKDRHFWRKSLRKL-329
MMP-3 hex2	445-AKKVTHTLK-453
MMP-3 cat	107-PKWRKTHL-114
MMP-1	107-PRWEQTHL-114
MMP-2	115-RKPKWDK-121
MMP-7	102-PKWTSKVV-109
MMP-8	106-PKWERTNL-113
MMP-9	104-LKWHHHNI-111
MMP-10	105-PKWRKTHL-112
MMP-11	102-GRWEKTDL-109
MMP-12	107-PVWRKH-112
MMP-13	111-LKWSKMNL-118
MMP-14	117-LKWQH-121
MMP-15	138-KW-139
MMP-16	126-KW-127
MMP-17	132-TKWNKRNL-139

Figure 2. Nuclear localization sequences found in human MMP proteins. The consensus sequence for the classical NLS is indicated at the top of the figure. Three NLSs from the MMP-3 pro-domain (pro1-3) and two NLSs from the hemopexin-like domain (hex1-2) are highlighted. The NLS from the catalytic domain of MMP-3 is shown as a reference sequence, for comparison purposes with putative NLSs from the catalytic domains of other MMPs. Similar or identical sequences within the aligned NLSs are indicated with blue.

For the first time, nTIMPs were reported in human gingival fibroblasts in 1995 and in human breast carcinoma cell lines in 1999, prior to the discovery of nMMPs [23,24]. Subsequently, Gasche et al. (2001), reported gelatinolytic activity in the nuclei of mouse brain cells after ischemia-reperfusion, for which nMMPs were suggested to be responsible for [65]. Two years later, Si-Tayeb et al. reported the detection of nMMP-3 in human hepatocellular carcinoma cell line (HepG2) and the identification of a nuclear localization signal (NLS) within the structure of the protease [66]. Since then, the number of reported nMMPs has grown, with some MMP members being localized to the nuclei in a variety of different cells types, originating from normal tissues, cancers, infected cells, and in cells during disease progression (Table 1) [67–71]. For example, nMMP-2 was found in normal skin cells in the lower one-third of the epidermis, whereas in the tumor and pre-cancerous samples, it was predominantly in the upper layers of the skin suggesting that the protein may be expressed at the early stages of squamous cell carcinogenesis [72]. The expression of MMP-7 and MMP-16 were also found in the nuclei of basal and supra-basal cells of normal squamous epithelium and condyloma [73]. Alternatively, MMP-12 was detected in the nuclei of the virus-transfected cells and MMP-14 was reported as being present in the nuclei of macrophages, supporting the possible involvement of MMPs in the immune response [74].

Table 1. Functions and localization of nuclear MMPs (nMMP) and nuclear TIMP (nTIMP). The table represents nMMPs and nTIMP1 and their functions in different cells and tissues. Malignant cells and tissues are indicated in red. Other pathological conditions are indicated in purple; a-deoxyribonucleic acid.

nMMP/nTIMP	Function	Cell Line or Tissue Type	Ref.
MMP-1	Apoptosis ↓	Human Muller glia	[75]
	Carcinogenesis ↑	Human breast cancer	[76]
	Not defined	Human keratinocytes, gingival tissue, megakaryocytes	[67,77,78]
MMP-2	Blood-brain barrier ↓	Mouse brain	[65]
	DNA[a] reparation ↓	Human mesothelioma, cardiac myocytes; rat liver; pig pulmonary artery endothelial cells	[34,55,79]
	DNA reparation ↓ Apoptosis ↑	Rat brain neurons	[80,81]
	Carcinogenesis ↑	Human hepatocellular carcinoma	[33]
	Muscle adaptation to training ↑	Rat skeletal muscle fibers	[82]
	Not defined	Human melanoma cells, cutaneous squamous cell carcinoma, actinic keratosis, normal skin, megakaryocytes, endothelial cells; rat neurons; mouse skeletal muscle fibers	[25,72,78,83–85]
MMP-3	Apoptosis ↑	Human hepatocellular carcinoma, hepatocellular carcinoma cell line, peritumoral liver, liver myofibroblasts; Chinese hamster ovary cells	[56,66]
	Cell migration ↑	Human normal, osteoarthritic chondrocytes	[58]
	Immune response ↑	Human embryonic kidney epithelial cell line, macrophages	[69]
	Not defined	Human megakaryocytes	[78]
MMP-7	Cell migration and wound healing ↑	Human prostate cancer cell lines; mouse prostate tumor	[68]
	Not defined	Human adenocarcinoma, condyloma, normal squamous, columnar epithelium	[73]
MMP-9	DNA reparation ↓ Apoptosis ↑	Rat brain neurons	[80,81]
	DNA reparation ↓	Human epithelioid mesothelioma cell line	[34]
	Osteoclastogenesis ↑	Mouse preosteoclasts	[86]
	Not defined	Human tubular atrophic renal tubules, gingival tissue, megakaryocytes; dog neuropil and neurons	[77,78,87,88]
MMP-12	Immune response ↑	Human cervical cancer cell line, myocardial cells, bronchial epithelial cell line, mouse fibroblasts, cardiomyocytes cell line	[70,89]
MMP-13	Carcinogenesis ↑	Human oral tongue squamous cell carcinoma	[36,90]
	Not defined	Human chondrosarcoma of the jaws, brain tissues; rat brain tissues, chondrocytes	[15,32,91]
MMP-14	Carcinogenesis ↑	Human hepatocellular carcinoma, hepatocellular carcinoma cell line	[33]
	Immune response ↑	Mouse bone marrow-derived macrophages	[74]
MMP-16	Not defined	Human adenocarcinoma, condyloma, normal squamous, columnar epithelium	[73]
TIMP1	Cell growth ↑	Human gingival fibroblasts cell line	[23]
	Not defined	Human breast carcinoma cell line, endothelial cells; rat neurons	[24,25]

The only nTIMP identified so far, nTIMP1, was found co-localized with nMMP-2 in endothelial cells and neurons, but no direct protein interactions or mechanism(s) for TIMPs translocation have been defined [25]. nTIMP1 inhibits nMMP-9, which was identified in neuronal cells, and exhibits insignificant or low levels of nMMP-9-derived gelatinase activity [80,81,87]. nTIMP1 was detected in gingival fibroblasts (in which nMMP-1 and nMMP-9 were later identified), and also in breast carcinoma cells, where nMMP-1 was also subsequently identified [23,24,76,77]. The mechanistic relationship shared between nMMPs and nTIMPs based upon them simultaneously residing in the nucleus have not been fully investigated in model systems, but immunofluorescence analysis has indicated nMMP/nTIMP co-localization within the nucleus of neural stem cells in Huntington's disease (HD) can contribute to enhanced neurotoxicity. Here, TGF-β treatment enhanced nTIMP1 protein levels, which conferred neuroprotection in HD against toxicity associated with the aggregation of neurotoxic mutant huntingtin proteins [71]. From Table 1, it is interesting to note that co-localization of nTIMP1 with every nMMP reported so far was detected (although not exclusively) in malignant tissues and a variety of other cell types. Additionally, while extracellular MMPs are seen to promote epithelial-mesenchymal transition (EMT), tumor invasion, and metastasis, nMMPs are reported to be present in both epithelial and the resulting mesenchymal cells, which is suggestive of them fulfilling potentially unique and distinct intracellular functions during EMT [92,93].

In summary, it is becoming firmly established that MMPs and TIMPs have the capacity to translocate to the cell's nucleus. While some of these depend on the presence of a classical NLS sequence (or a sequence derived from this), others are capable of this event through caveolin-dependent endocytosis. Unveiling the underlying molecular mechanisms for nMMP and nTIMP transport may lay down solid foundations for such mechanisms to be potentially targeted.

3. Nuclear MMPs as Regulators of Gene Expression

Following the entry of MMPs into the nucleus, they have been shown to participate in a number of different processes, such as cell migration, proliferation, signaling pathways, tumor growth, and the immune response (Figure 3) [60,68,75,94–96]. As an area of research that has seen significant growth over the recent years, we outline a number of key publications that highlight the diverse biological effects that are modulated when the MMPs are resident within the nucleus.

Unlike the extracellular MMPs, nMMPs have access to genomic DNA and may therefore modulate gene expression events related to disease progression. For example, in the human bone osteosarcoma epithelial cells, MMP-2 was visualized by immunofluorescence methods in the nucleolus where it could interact with DNA associated with different regions of the ribosomal RNA genes, suggestive of its potential to regulate rRNA transcriptional initiation [97]. Here, the inhibition of MMP-2 activity by siRNA interference led to a slower cell proliferation rate in comparison to control cells.

In human breast carcinoma MCF7 cells, overexpression of MMP-14 significantly increased the transcriptional expression of vascular endothelial growth factor A (VEGF-A) [95]. Mechanistically, MMP-14-regulated VEGF-A expression could be suppressed through the treatment of cells with the Src-tyrosine kinase inhibitor PP2 [95], and whether this regulatory effect is direct or not is still to be revealed. Similarly, such findings also have great significance for the role of nMMP-14 in the promotion of tumor growth or invasiveness [98,99]. Here, nMMP-14 stimulated the expression of SMAD1 via TGF-β signaling [98]. Additionally, nMMP-14 suppressed the expression of Dickkopf-3 (DKK3) in human urothelial cell carcinoma tissue, which led to increased invasiveness of cells [99]. In support of this, while the localization of nMMP-14 was not the object of the investigation, the nuclear staining of MMP-14 has also been observed and reported independently by immunohistochemical methods in other studies [33,74].

Immunocytochemistry methods have also identified nMMP-3 [58]. It was shown that the HEX domain of nMMP-3 can interact with transcription enhancer dominant in chondrocytes (TRENDIC) within the connective tissue growth factor gene (*CTGF/CCN2*) promoter region and activate its transcription [58,60]. The proteins regulated by this promoter play an important role in proliferation, the formation of the extracellular matrix, angiogenesis, and cell migration. In human dental pulp, nuclear MMP-3 could also regulate the expression of CTGF/CCN2 proteins and the cellular migration capacity of cells through this pathway [60].

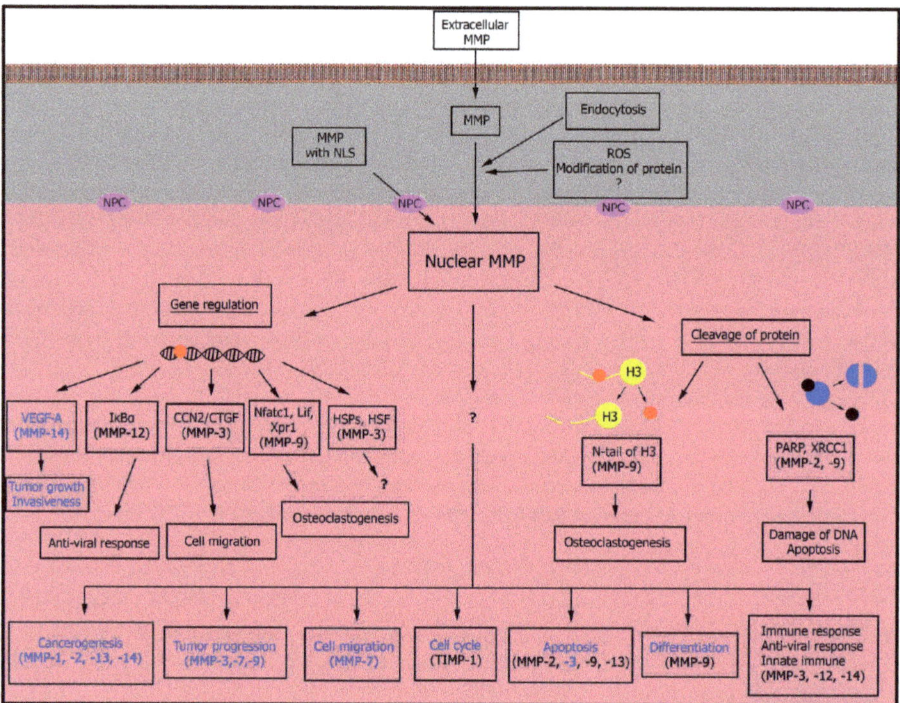

Figure 3. Role of nuclear MMPs within cells. Currently, only two mechanisms for the transport of MMPs into the cell's nucleus are known, and which utilize the nuclear pore complex or endocytosis. Within the nucleus, MMPs can cleave nuclear proteins or regulate the transcription of various genes. Through these mechanisms, nMMPs can modulate a number of key biological processes within the cell. Participation of nMMPs in cancer cell progression is indicated in blue. NPC—nuclear pore complex; MMP—matrix metalloproteinase; NLS—nuclear localization signal.

Although nuclear MMPs are known to participate in malignant tumor progression, they also have additional functions of importance that are related to normal cellular homeostasis. Here, the use of transcriptomic analyses revealed that overexpression of MMP-3 stimulated mRNA expression of heat shock proteins (HSPs), HSP70B, HSP72, HSP40, and HSP20. Several transcription factors that potentially interact with nMMP-3 were predicted and one of them, heat shock factor 1 (HSF1) was validated to co-activate the *HSP70B* gene promoter together with the nMMP-3 protein [59]. Of note, the HEX domain alone was sufficient to induce HSP70B expression. Other transcriptional factors that nMMP-3 may associate with include FOXO3, VDR, Ets-1, CULT1, TBP, and SP1. Since MMP-3-green fluorescent protein (GFP) was found in cellular chromatin fractions and soluble nuclear fractions of COS7 cells, in which nuclear markers chromobox protein CBX5/HP1α and histone-H3 were also detected, it was suggested that MMP-3 can also enter the cell's nucleus to possibly modulate gene expression events [59].

Another protease, MMP-9, was found to contribute to osteoporosis, which is characterized by increased osteoclastogenesis and a decreased number of active osteoblasts (for bone formation). During osteoclastogenesis, nMMP-9 affected the expression of more than 67% of genes [86], normally expressed in primary osteoclast precursor cells, which included genes that regulate RANKL, AMPK, and VEGF signaling pathways. On a morphological level, inhibition of nMMP-9 enzyme activity led to reduced maturation of osteoclasts, compared with control cells. Using an alternative approach by incorporating ChiPac-seq technology, nMMP-9 was seen as being required for histone-H3 protein cleavage near the transcription start sites of the osteoclastogenic genes *Nfatc1*, *Lif*, *Xpr1*, and for their concurrent activation during osteoclastogenesis. Nuclear accumulation of MMP-9 was also confirmed by immunofluorescence microscopy [86].

Tetracyclines have antimicrobial activity, block bone deterioration and work as MMP-9 inhibitors [99]. Tetracycline analogs, minocycline, and tigecycline suppressed osteoclast formation by blocking nMMP-9-mediated proteolysis of the amino-terminal of histone-H3 protein [100]. Antibiotic treatments significantly reduced the differentiation of osteoclasts but did not affect the proliferation of pre-osteoblasts and osteoclast precursor cells. At the transcriptional level, both tetracycline analogs repressed RANKL-induced mRNA expression of the MMP-9-targeted genes *Nfatc1*, *Lif*, and *Xpr1*. Through using tigecycline and minocycline treatments on zebrafish larvae harboring an osteoporosis phenotype, the antibiotics reduced prednisolone-induced osteoporosis in a dose-dependent manner. Such antibiotics can therefore be effective as a treatment for osteoporosis through modulating nMMP-9 enzymatic activity towards the histone-H3 protein and its gene regulatory effects [100].

In summary, a number of novel functions for nuclear MMPs have emerged over the years, spanning the mechanistic entry of MMPs to the nucleus and their input into the regulation of gene expression in cell- and disease-context dependent manner. In particular, some nMMPs affect proliferation, migration, and invasion that contribute to tumor progression. These data also support the fact that nMMPs can influence cellular processes at the level of gene regulation, thus highlighting additional potential as targets in the treatment of cancer.

4. Nuclear MMPs as Regulators of Malignancy

Nuclear MMPs also mediate a malignant cell phenotype while contributing to cellular mobility and tumor progression. For example, nMMP-7 together with alternative reading frame (ARF) protein expression contributed to enhanced migration and metastasis of prostate cancer cells [68]. Knockdown of ARF expression in cancer cells decreased MMP-7 expression, but when ARF was over-expressed, MMP-7 accumulated in the nucleus where it could bind to the ARF protein. The molecular mechanisms responsible for this effect remain unclear, but the concurrent increase in these two proteins within the nucleus is correlated with malignancy of cancer cells and the combined targeting of ARF and MMP-7 may therefore have therapeutic value in the treatment of advanced prostate cancer.

The proteases MMP-3 and MMP-9 also contribute to tumor progression [35]. Significant expression of both non-proteolytically-active and proteolytically-active isoforms of these MMPs were found in metastatic cells derived from colon adenocarcinoma cell lines. After colon adenocarcinoma cells were injected into the abdominal walls of mice, primary tumors and metastatic tumors in lung tissues contained active nMMP-9 that had become localized within the nuclei of cells detected within the tumor-stromal area. The knock-down expression of MMP3 by siRNA during the latter experiments suppressed cancer cell migration, suggesting an important and significant contribution from nMMP-9 and MMP-3 during tumor invasiveness [35].

In summary, it is becoming increasingly apparent that intracellular MMPs can play multiple (yet significant) roles in tumor progression.

5. Nuclear MMPs and Oxidative Damage to DNA

The MMP proteases can process multiple DNA-interacting nuclear proteins during oxidative stress. For example, the accumulation and activation of MMPs were observed in the nuclei of ischemic cells after reperfusion [80,101]. Here, MMP-14 promoted the activation of the zymogens pro-MMP-2 and pro-MMP-9 within the nuclei of ischemic cells after reperfusion [80]. Catalytically active nMMP-2 and nMMP-9 have been shown to cleave the PARP-1 and XRCC1 proteins, which play an important role in DNA repair and caspase-independent cellular apoptosis [44,55]. It was reported that adenosine diphosphate could enhance the cleavage of PARP-1 by nMMP-2 and nMMP-9, through the PI3K/Akt/NF-κB and ERK1/2 signal transduction pathways [34]. Cleavage of PARP-1 and XRCC1 was also observed, which led to the accumulation of damaged DNA within the nuclei of ischemic brain cells after reperfusion. When rats were treated with the broad-spectrum MMP inhibitor BB1101, the cleavage of PARP-1 was significantly reduced and the amounts of XRCC1 protein were reported to increase. The use of such inhibitors in therapeutically targeting MMPs following cerebral ischemia-reperfusion injury is a good example of how nMMPs are being targeted for therapeutic purposes [81].

Other studies have also confirmed the accumulation of MMP-2 and MMP-9 in the nuclei upon ischemia treatment of cells. In the nuclei of mouse neurons deficient in superoxide dismutase (SOD1) and treated with ischemia-reperfusion, pro-MMP-2 and pro-MMP-9 protein levels were induced and activated. Active MMP-2 and MMP-9 are involved in the early-stage destruction of the blood-brain barrier, caused by oxidative stress during cerebral ischemia-reperfusion [65]. After an ischemic stroke in neurons and glial cells, protein MMP-9 was reported to be localized in the nucleus. The cells containing nMMP-9 also expressed activated caspase 3, which confirmed the link between the nuclear localization of MMP-9 and neuronal apoptosis in ischemic cells [102]. Similarly, the activated form of MMP-13 was also found in the nuclei of neurons as an early event following cerebral ischemia [15]. By subjecting the primary neural culture of rats to oxygen and glucose deprivation, Cuadrado et al. (2009) were able to demonstrate the nuclear translocation of MMP-13 in vitro.

Collectively, such important findings suggest that nMMPs may also fulfill a role in modulating the cell's response to oxidative stress (in addition to disease progression) and that certain members of this family may also be directly be involved in the DNA damage response and caspase-dependent cell death.

6. Nuclear MMP and Apoptosis

Regulated cell death can occur in different ways in the form of apoptosis, necrosis, pyroptosis, and autophagy [103]. Whether the cell chooses the path of apoptosis or the path of survival depends on the ratio of pro- and anti-apoptotic factors, and in this context, MMPs have been found to modulate both pro-apoptotic [104] and pro-survival effects [105].

One of the environmental factors that can induce oxidative stress and apoptosis is cigarette smoke, which reportedly changes the expression levels of MMP-2, MMP-9, and TIMP-2 and their subcellular localization in pulmonary artery endothelial cells [79]. Here, the level of annexin V-positive/propidium iodide-negative cells significantly increased compared to untreated control cells indicative of enhanced apoptosis. Cells exposed to cigarette smoke contained PARP-1 protein fragments usually detected in apoptotic cells, including a high level of gelatinase activity. Since MMP-2 and MMP-9 were also observed to cleave PARP-1, these data suggest that cigarette smoke may induce apoptosis via MMP activation.

Alternatively, nMMP-1 may also have a pro-survival role. For example, MMP-1 is co-localized with mitochondria and the nucleus in normal glial Muller cells [75], but during staurosporine-induced apoptosis, MMP-1 expression changes and localizes to perinuclear mitochondrial clusters and around fragmented nuclei. Inhibition of MMP-1 activity led to lamin degradation, caspases activation, and apoptosis.

Nuclear MMP-3 expression in HepG2 and liver myofibroblast cells could also affect their rate of apoptosis [56]. When Chinese hamster ovary cells were transfected by a plasmid encoding an EFGP/active MMP-3 fusion protein, it principally localized in the nuclei. Using an antibody against activated caspase 3, it was determined that cells transfected with EGFP/active MMP-3 had higher apoptotic levels compared with untransfected cells. This effect was enhanced in cells where MMP-3 was present within the nucleus. Moreover, expression of a catalytically-inactive form of MMP-3 or inhibition of wild type MMP-3 in the presence of a broad-spectrum MMP inhibitor GM6001, led to a reduction in apoptotic cells. Such findings suggest another important biological effect for active nMMP-3 in apoptosis regulation.

Collectively, nuclear MMPs can have the effects of enhancing or decreasing apoptosis of cells. As described above, there is a clear relationship between nuclear MMP-2, MMP-3, MMP-9, and activation of apoptosis. Conversely, nMMP-1 can block the pathway of apoptosis. Such findings have been defined for a limited number of the MMP family members and clearly further developments in this area of research are warranted based on the importance of nMMPs and their contribution to disease progression [106].

7. Nuclear MMPs in Immune and Anti-Viral Responses

During inflammation, the expression profiles and activity of a wide range of proteases is increased. These include serine proteinases, such as granzymes, neutrophilic elastases, cathepsin G, and proteinase 3 [107]. Some of these proteases can modulate inflammation and the immune response via regulation of cytokines and chemokines [108]. For example, innate immunity is regulated by MMP-25, which is preferentially produced by leukocyte cells. While MMP-25-deficient mice were viable, they had defects in their innate immune system through having high sensitivity to bacterial lipopolysaccharide, hypergammaglobulinemia, and showed decreased secretion of the pro-inflammatory molecule COX2 [109]. In macrophages, nMMP-14 was observed to participate in the regulation of inflammation, the immune response, or anti-viral and innate immunity. Mechanistically, nMMP-14 can trigger expression and activate phosphoinositide 3-kinase δ-Akt-GSK3β signal cascade and modulate the Mi-2/NuRD nucleosome remodeling complex [74].

The immune system is also modulated by MMPs regulating the transcription of genes involved in anti-viral immunity. For example, macrophages can release MMP-12 during a viral infection, which can also enter infected cells and be translocated to the nucleus, presumably through endocytosis and lipid-dependent trafficking [95]. Through its catalytic domain, nMMP-12 can bind the polyA-rich regions within the promoter region of the *IκBα* encoding gene, and induce its transcriptional expression [70,89], which upregulates the secretion of interferon-alpha (IFN-α) [70]. However, in the absence of MMP-12 expression (in a knockout mouse model), in mice infected with Coxsackie B type B2 virus, unsecreted IFN-α protein remained within pancreatic, heart, and hepatocyte cells and the mice succumbed to the lethal effects of viral infection. Such effects on IFN-α expression could be reversed upon the artificial expression of MMP-12 in MMP-12$^{-/-}$ fibroblast cells in vitro [70]. Moreover, nMMP-12 expression was reported to reside in the nucleus of human cardiomyocyte cells [70] and exogenously added recombinant MMP-12 protein, or its catalytic domain, observed to traffic to the nucleus, when used to treat MMP-12-silenced HeLa cells. Additionally, extracellular MMP-12 could cleave and inactivate systemic IFN-α, thereby attenuating the anti-viral inflammatory response as part of a negative feedback loop and which could be reversed upon treating virally-infected mice with the MMP-12 inhibitor, RXP470. Here, morbidity in mice was observed to be reduced, as was viral replication. Collectively, inhibition of extracellular MMP-12 or increasing its nuclear localization (or activity) highlights a potential basis on which the development of a therapeutic strategy against viral infection could be implemented.

The cellular anti-viral response against Dengue virus is augmented by MMP-3 [69]. Zuo et al. observed that the presence of nMMP-3 within infected cells was increased. The silencing of MMP-3 led to increased titers of the virus, decreased levels of cytokines and chemokines, and the reduced activity of NF-κB. Since nMMP-3 was found to be co-localized with intracellular NF-κB, it was suggested that the protease up-regulated the activity of NF-κB via a direct protein-protein interaction, which could subsequently promote the transcription of anti-viral and pro-inflammatory genes [69]. Collectively, such findings suggest that nMMP-3 plays a significant role in the anti-viral defense of the body against Dengue virus.

The requirement for MMPs for the immune response and protection of the organism against various infections has been suggested previously [110]. Predominantly, extracellular MMPs regulate the migration of immune cells, proteolysis of the basement membrane and the remodeling of the extracellular matrix. In the nucleus, MMPs are emerging to participate in the regulation of gene expression and regulate immunity against viruses and bacteria [69,70,74,111], but their additional abilities to cleave gene products that are central to negatively regulating the immune response cannot be completely excluded at this juncture.

8. Future Directions

MMPs were first identified in 1962 and since then have been characterized as extracellular proteases [112] which are firmly established as playing critical roles in oncogenesis and other pathological processes [113]. Over the years, MMPs have been understood as being some of the most pursued targets for drug development [9]. Their large family size, redundant roles, and substrate specificity are good reasons for why side effects arising from targeting them during disease progression with novel therapeutics has been a major obstacle for good therapeutics reaching the clinic. However, the detection of MMPs within nucleus, when taken with the biological effects they regulate, raises renewed optimism for developing therapeutics that are specific for these MMP derivatives. Mechanistically, the translocation of MMPs to the nucleus have only been thoroughly investigated for MMP-3 [58]. While bioinformatics approaches enable the identification of putative NLSs in other MMPs, their ability to translocate the proteases into nucleus and the mechanisms they utilize to do this still remain to be unveiled [61]. Such investigations would help in defining the plausibility of targeting nMMPs with the foresight of minimizing unwanted side effects, with greater clarity. The activity of MMPs can also be regulated by TIMPs [19]. So far, only TIMP1 has been reported within nuclei and no mechanism of its translocation has been described [25]. Since TIMPs also inhibit other a disintegrin and metalloproteinases (ADAMs), such as ADAM-10, therapeutic targeting nTIMP (so that nMMPs can take greater effect) may offer limitations [114].

A number of approaches have been adopted with a view to targeting MMPs for therapeutic purposes. For example, the small-molecule inhibitors hydroxamic acid, carboxylic acid, 5,5-disubstituted barbiturates, benzosulfonamide, and phosphonate have all shown efficacy in reducing oncogenesis, but unwanted side-effects have presented a number of challenges [115]. Alternatively, the more specific approach of targeting metalloproteases using single chain antibody fragments (scFV) has shown some encouraging outcomes in vitro [116–118]. Similarly, scFv fragments developed against extracellular MMP-14 have also shown good efficacy against cancer cell invasiveness in cell line models validated in a mouse orthotopic xenograft model [118]. Moreover, monoclonal antibodies directed at MMP-14 also successfully prevented the activation of pro-MMP-2 while antibodies to MMP-9 interfered with the catabolism of gelatin [119]. Collectively, while such approaches do encouragingly highlight the feasibility of targeting extracellular MMPs, their ability to target nMMPs remain to be explored.

For effective nMMP-specific inhibitor design, it may be necessary to elucidate the biological functions of distinct nMMPs and their mechanisms utilized for nuclear translocation. Since some of the proteases play important physiological and biological roles, as in the immune and anti-viral response, while others even suppress tumor growth via apoptosis, inhibiting nMMPs may require careful consideration that would potentially leave otherwise favorable biological effects intact [56,66,69,70]. Nevertheless, one serious challenge in targeting nMMPs is the specific delivery of the inhibitor to the nucleus and over the last ten years, several nano-carriers have been developed to help overcome this potential obstacle [120]. Such carriers have proven their efficiency in drug-targeting approaches for human cervical cancer, human oral squamous carcinoma cell lines, and multidrug-resistant breast cancer cell lines and in vivo, using MCF-7-derived breast tumor-bearing mice [121–124]. Alternatively, the use of such carrier systems with MMP inhibitors in combination with other conventional therapeutic reagents may have some usefulness to help combat tumor development, migration and metastasis potential [125–128].

9. Conclusions

Extracellular MMPs regulate a variety of functions, such as the development of tissues, inflammation, apoptosis, migration, angiogenesis, vasculogenesis, and other processes. However, it is emerging that nuclear matrix metalloproteinases are functionally distinct, through them performing unique and mutually exclusive functions within the nucleus. Although almost 25 years have passed since the first reports of nuclear localization of MMPs appeared, much remains to be explored. Here, one of these key areas is how the matrix metalloproteinases, whether secreted or anchored in the cell membrane, are transported to the cell's nucleus. While some MMPs encode a classical NLS, others are transported through endocytosis. The activity of nMMPs is diverse in that they can promote tumor metastasis and other pathological processes. While on the one hand, nMMPs can contribute to apoptosis resulting in tumor cell death, on the other hand, nMMPs can positively regulate the immune response towards viral and bacterial infections. Surprisingly, the MMP inhibitor TIMP1 has also been detected within the nucleus and its full repertoire of inhibitory functions remains to be fully elucidated, in addition to whether other TIMPs can also reside in the nucleus.

Depending on the favorable or unfavorable effects of nMMP proteins, there appears to be some flexibility presented in how nMMPs may be targeted based on the manner in which they mechanistically translocate to the nucleus. For example, one can try to respectively elevate or interrupt nuclear transport through targeting nTIMP-chaperone effects or directly targeting the nMMPs in a "compartment-specific" manner as the proteases traffic to the nucleus. To create specific targeting approaches for nMMPs activities, it is necessary to understand the biochemical network of these proteases in detail and gain a greater understanding of what other key biological effects these proteases may be regulating during disease progression, as a fundamental prerequisite.

Author Contributions: Conceptualization, A.S.F., S.M.S. and A.A.Z.J.; writing—original draft preparation, A.S.F., A.I.P. and S.M.S.; writing—review and editing, A.S.F., A.I.P., V.A.M. and S.M.S.; visualization, A.S.F. and A.I.P.; supervision, S.M.S. and A.A.Z.J.; funding acquisition, A.A.Z.J. All authors have read and agreed to the published version of the manuscript.

Funding: This research was funded by the Russian Science Foundation, grant number 16-15-10410.

Conflicts of Interest: The authors declare no conflict of interest.

References

1. Cui, N.; Hu, M.; Khalil, R.A. Biochemical and Biological Attributes of Matrix Metalloproteinases. In *Progress in Molecular Biology and Translational Science*; Elsevier B.V.: Amsterdam, The Netherlands, 2017; Volume 147, pp. 1–73.
2. Houghton, A.M.G.; Grisolano, J.L.; Baumann, M.L.; Kobayashi, D.K.; Hautamaki, R.D.; Nehring, L.C.; Cornelius, L.A.; Shapiro, S.D. Macrophage elastase (matrix metalloproteinase-12) suppresses growth of lung metastases. *Cancer Res.* **2006**, *66*, 6149–6155. [CrossRef] [PubMed]
3. Kobayashi, T.; Kim, H.; Liu, X.; Sugiura, H.; Kohyama, T.; Fang, Q.; Wen, F.-Q.; Abe, S.; Wang, X.; Atkinson, J.J.; et al. Matrix metalloproteinase-9 activates TGF-β and stimulates fibroblast contraction of collagen gels. *Am. J. Physiol. Cell. Mol. Physiol.* **2014**, *306*, L1006–L1015. [CrossRef] [PubMed]
4. Chun, T.H.; Hotary, K.B.; Sabeh, F.; Saltiel, A.R.; Allen, E.D.; Weiss, S.J. A Pericellular Collagenase Directs the 3-Dimensional Development of White Adipose Tissue. *Cell* **2006**, *125*, 577–591. [CrossRef]
5. Caley, M.P.; Martins, V.L.C.; O'Toole, E.A. Metalloproteinases and Wound Healing. *Adv. Wound Care* **2015**, *4*, 225–234. [CrossRef] [PubMed]
6. Page-McCaw, A.; Ewald, A.J.; Werb, Z. Matrix metalloproteinases and the regulation of tissue remodelling. *Nat. Rev. Mol. Cell Biol.* **2007**, *8*, 221–233. [CrossRef]
7. Burrage, P.S. Matrix Metalloproteinases: Role in Arthritis. *Front. Biosci.* **2006**, *11*, 529. [CrossRef] [PubMed]
8. Papazafiropoulou, A.; Tentolouris, N. Matrix metalloproteinases and cardiovascular diseases. *Hippokratia* **2009**, *13*, 76–82.

9. Jabłońska-Trypuć, A.; Matejczyk, M.; Rosochacki, S. Matrix metalloproteinases (MMPs), the main extracellular matrix (ECM) enzymes in collagen degradation, as a target for anticancer drugs. *J. Enzyme Inhib. Med. Chem.* **2016**, *31*, 177–183. [CrossRef]

10. Rawlings, N.D.; Barrett, A.J.; Thomas, P.D.; Huang, X.; Bateman, A.; Finn, R.D. The MEROPS database of proteolytic enzymes, their substrates and inhibitors in 2017 and a comparison with peptidases in the PANTHER database. *Nucleic Acids Res.* **2018**, *46*, D624–D632. [CrossRef]

11. Rhizobium, G.E. Complete Genome Sequence of the Sesbania Symbiont and Rice. *Nucleic Acids Res.* **2013**, *1*, 13–14.

12. Nagase, H.; Visse, R.; Murphy, G. Structure and function of matrix metalloproteinases and TIMPs. *Cardiovasc. Res.* **2006**, *69*, 562–573. [CrossRef]

13. Chakraborti, S.; Mandal, M.; Das, S.; Mandal, A.; Chakraborti, T. Regulation of matrix metalloproteinases. An overview. *Mol. Cell. Biochem.* **2003**, *253*, 269–285. [CrossRef] [PubMed]

14. Roghi, C.; Jones, L.; Gratian, M.; English, W.R.; Murphy, G. Golgi reassembly stacking protein 55 interacts with membrane-type (MT) 1-matrix metalloprotease (MMP) and furin and plays a role in the activation of the MT1-MMP zymogen. *FEBS J.* **2010**, *277*, 3158–3175. [CrossRef] [PubMed]

15. Cuadrado, E.; Rosell, A.; Borrell-Pagès, M.; García-Bonilla, L.; Hernández-Guillamon, M.; Ortega-Aznar, A.; Montaner, J. Matrix metalloproteinase-13 is activated and is found in the nucleus of neural cells after cerebral ischemia. *J. Cereb. Blood Flow Metab.* **2009**, *29*, 398–410. [CrossRef] [PubMed]

16. Maquoi, E.; Frankenne, F.; Baramova, E.; Munaut, C.; Sounni, N.E.; Remacle, A.; Murphy, G.; Foidart, J. Membrane Type 1 Matrix Metalloproteinase-associated Degradation of Tissue Inhibitor of Metalloproteinase 2 in Human Tumor Cell Lines. *J. Biol. Chem.* **2000**, *275*, 11368–11378. [CrossRef] [PubMed]

17. Madzharova, E.; Kastl, P.; Sabino, F.; auf dem Keller, U. Post-Translational Modification-Dependent Activity of Matrix Metalloproteinases. *Int. J. Mol. Sci.* **2019**, *20*, 3077. [CrossRef] [PubMed]

18. Petushkova, A.I.; Zamyatnin, A.A. Redox-mediated post-translational modifications of proteolytic enzymes and their role in protease functioning. *Biomolecules* **2020**, *10*, 650. [CrossRef]

19. Jackson, H.W.; Defamie, V.; Waterhouse, P.; Khokha, R. TIMPs: Versatile extracellular regulators in cancer. *Nat. Rev. Cancer* **2017**, *17*, 38–53. [CrossRef]

20. Batra, J.; Soares, A.S.; Mehner, C.; Radisky, E.S. Matrix Metalloproteinase-10/TIMP-2 Structure and Analyses Define Conserved Core Interactions and Diverse Exosite Interactions in MMP/TIMP Complexes. *PLoS ONE* **2013**, *8*, e75836. [CrossRef]

21. Wisniewska, M.; Goettig, P.; Maskos, K.; Belouski, E.; Winters, D.; Hecht, R.; Black, R.; Bode, W. Structural Determinants of the ADAM Inhibition by TIMP-3: Crystal Structure of the TACE-N-TIMP-3 Complex. *J. Mol. Biol.* **2008**, *381*, 1307–1319. [CrossRef]

22. Arpino, V.; Brock, M.; Gill, S.E. The role of TIMPs in regulation of extracellular matrix proteolysis. *Matrix Biol.* **2015**, *44–46*, 247–254. [CrossRef] [PubMed]

23. Li, H.; Nishio, K.; Yamashita, K.; Hayakawa, T.; Hoshino, T. Cell cycle-dependent localization of tissue inhibitor of metalloproteinases-1 immunoreactivity in cultured human gingival fibroblasts. *Nagoya J. Med. Sci.* **1995**, *58*, 133–142. [CrossRef] [PubMed]

24. Ritter, L.M.; Garfield, S.H.; Thorgeirsson, U.P. Tissue inhibitor of metalloproteinases-1 (TIMP-1) binds to the cell surface and translocates to the nucleus of human MCF-7 breast carcinoma cells. *Biochem. Biophys. Res. Commun.* **1999**, *257*, 494–499. [CrossRef] [PubMed]

25. Sinha, S.K.; Asotra, K.; Uzui, H.; Nagwani, S.; Mishra, V.; Rajavashisth, T.B. Nuclear localization of catalytically active MMP-2 in endothelial cells and neurons. *Am. J. Transl. Res.* **2014**, *6*, 155–162.

26. Goffin, L.; Fagagnini, S.; Vicari, A.; Mamie, C.; Melhem, H.; Weder, B.; Lutz, C.; Lang, S.; Scharl, M.; Rogler, G.; et al. Anti-MMP-9 Antibody: A Promising Therapeutic Strategy for Treatment of Inflammatory Bowel Disease Complications with Fibrosis. *Inflamm. Bowel Dis.* **2016**, *22*, 2041–2057. [CrossRef]

27. Chaturvedi, M.; Kaczmarek, L. MMP-9 inhibition: A therapeutic strategy in ischemic stroke. *Mol. Neurobiol.* **2014**, *49*, 563–573. [CrossRef]

28. Vandenbroucke, R.E.; Dejonckheere, E.; Libert, C. Series "Matrix metalloproteinases in lung health and disease": A therapeutic role for matrix metalloproteinase inhibitors in lung diseases? *Eur. Respir. J.* **2011**, *38*, 1200–1214. [CrossRef]

29. Sawicki, G. Intracellular Regulation of Matrix Metalloproteinase-2 Activity: New Strategies in Treatment and Protection of Heart Subjected to Oxidative Stress. *Scientifica* **2013**, *2013*, 130451. [CrossRef]

30. Fields, G.B. Mechanisms of action of novel drugs targeting angiogenesis-promoting matrix metalloproteinases. *Front. Immunol.* **2019**, *10*, 1278. [CrossRef]

31. Levin, M.; Udi, Y.; Solomonov, I.; Sagi, I. Next generation matrix metalloproteinase inhibitors—Novel strategies bring new prospects. *Biochim. Biophys. Acta Mol. Cell Res.* **2017**, *1864*, 1927–1939. [CrossRef]

32. Zyada, M.M.; Shamaa, A.A. Is collagenase-3 (MMP-13) expression in chondrosarcoma of the jaws a true marker for tumor aggressiveness? *Diagn. Pathol.* **2008**, *3*, 26. [CrossRef] [PubMed]

33. Ip, Y.C.; Cheung, S.T.; Fan, S.T. Atypical localization of membrane type 1-matrix metalloproteinase in the nucleus is associated with aggressive features of hepatocellular carcinoma. *Mol. Carcinog.* **2007**, *45*, 225–230. [CrossRef] [PubMed]

34. Muscella, A.; Cossa, L.G.; Vetrugno, C.; Antonaci, G.; Marsigliante, S. Adenosine diphosphate regulates MMP2 and MMP9 activity in malignant mesothelioma cells. *Ann. N. Y. Acad. Sci.* **2018**, *1431*, 72–84. [CrossRef] [PubMed]

35. Okusha, Y.; Eguchi, T.; Sogawa, C.; Okui, T.; Nakano, K.; Okamoto, K.; Kozaki, K.I. The intranuclear PEX domain of MMP involves proliferation, migration, and metastasis of aggressive adenocarcinoma cells. *J. Cell. Biochem.* **2018**, *119*, 7363–7376. [CrossRef] [PubMed]

36. Mäkinen, L.K.; Häyry, V.; Atula, T.; Haglund, C.; Keski-Säntti, H.; Leivo, I.; Mäkitie, A.; Passador-Santos, F.; Böckelman, C.; Salo, T.; et al. Prognostic significance of matrix metalloproteinase-2, -8, -9, and -13 in oral tongue cancer. *J. Oral Pathol. Med.* **2012**, *41*, 394–399. [CrossRef] [PubMed]

37. Puchi, M.; García-Huidobro, J.; Cordova, C.; Aguilar, R.; Dufey, E.; Imschenetzky, M.; Bustos, P.; Morin, V. A new nuclear protease with cathepsin L properties is present in HeLa and Caco-2 cells. *J. Cell. Biochem.* **2010**, *111*, 1099–1106. [CrossRef]

38. Soond, S.M.; Kozhevnikova, M.V.; Frolova, A.S.; Savvateeva, L.V.; Plotnikov, E.Y.; Townsend, P.A.; Han, Y.P.; Zamyatnin, A.A. Lost or Forgotten: The nuclear cathepsin protein isoforms in cancer. *Cancer Lett.* **2019**, *462*, 43–50. [CrossRef]

39. Bach, A.S.; Derocq, D.; Matha, V.L.; Montcourrier, P.; Sebti, S.; Orsetti, B.; Theillet, C.; Gongora, C.; Pattingre, S.; Ibing, E.; et al. Nuclear cathepsin D enhances TRPS1 transcriptional repressor function to regulate cell cycle progression and transformation in human breast cancer cells. *Oncotarget* **2015**, *6*, 28084–28103. [CrossRef]

40. Wang, X.; Sato, R.; Brown, M.S.; Hua, X.; Goldstein, J.L. SREBP-1, a membrane-bound transcription factor released by sterol-regulated proteolysis. *Cell* **1994**, *77*, 53–62. [CrossRef]

41. Gourdet, C.; Iribarren, C.; Morin, V.; Bustos, P.; Puchi, M.; Imschenetzky, M. Nuclear cysteine-protease involved in male chromatin remodeling after fertilization is ubiquitously distributed during sea urchin development. *J. Cell. Biochem.* **2007**, *101*, 1–8. [CrossRef]

42. Paroni, G.; Henderson, C.; Schneider, C.; Brancolini, C. Caspase-2 can trigger cytochrome c release and apoptosis from the nucleus. *J. Biol. Chem.* **2002**, *277*, 15147–15161. [CrossRef] [PubMed]

43. Blagosklonny, M.V.; An, W.G.; Melillo, G.; Nguyen, P.; Trepel, J.B.; Neckers, L.M. Regulation of BRCA1 by protein degradation. *Oncogene* **1999**, *18*, 6460–6468. [CrossRef] [PubMed]

44. Ning, W.; Dong, Y.; Sun, J.; Li, C.; Matthay, M.A.; Feghali-Bostwick, C.A.; Choi, A.M.K. Cigarette smoke stimulates matrix metalloproteinase-2 activity via EGR-1 in human lung fibroblasts. *Am. J. Respir. Cell Mol. Biol.* **2007**, *36*, 480–490. [CrossRef] [PubMed]

45. Cauwe, B.; Opdenakker, G. Intracellular substrate cleavage: A novel dimension in the biochemistry, biology and pathology of matrix metalloproteinases. *Crit. Rev. Biochem. Mol. Biol.* **2010**, *45*, 351–423. [CrossRef] [PubMed]

46. Christensen, S.; Purslow, P.P. The role of matrix metalloproteinases in muscle and adipose tissue development and meat quality: A review. *Meat Sci.* **2016**, *119*, 138–146. [CrossRef]

47. Jobin, P.G.; Butler, G.S.; Overall, C.M. New intracellular activities of matrix metalloproteinases shine in the moonlight. *Biochim. Biophys. Acta Mol. Cell Res.* **2017**, *1864*, 2043–2055. [CrossRef]

48. Mannello, F.; Medda, V. Nuclear localization of matrix metalloproteinases. *Prog. Histochem. Cytochem.* **2012**, *47*, 27–58.

49. Siemianowicz, K.; Likus, W.; Markowski, J. Metalloproteinases in Brain Tumors. In *Molecular Considerations and Evolving Surgical Management Issues in the Treatment of Patients with a Brain Tumor*; IntechOpen: London, UK, 2015. [CrossRef]

50. Xie, Y.; Mustafa, A.; Yerzhan, A.; Merzhakupova, D.; Yerlan, P.; Orakov, A.N.; Wang, X. Nuclear matrix metalloproteinases: Functions resemble the evolution from the intracellular to the extracellular compartment. *Cell Death Discov.* **2017**, *3*, 17036. [CrossRef]

51. Hadler-Olsen, E.; Fadnes, B.; Sylte, I.; Uhlin-Hansen, L.; Winberg, J.O. Regulation of matrix metalloproteinase activity in health and disease. *FEBS J.* **2011**, *278*, 28–45. [CrossRef]

52. Gonzalez-Avila, G.; Sommer, B.; Mendoza-Posada, D.A.; Ramos, C.; Garcia-Hernandez, A.A.; Falfan-Valencia, R. Matrix metalloproteinases participation in the metastatic process and their diagnostic and therapeutic applications in cancer. *Crit. Rev. Oncol. Hematol.* **2019**, *137*, 57–83. [CrossRef]

53. Lange, A.; Mills, R.E.; Lange, C.J.; Stewart, M.; Devine, S.E.; Corbett, A.H. Classical nuclear localization signals: Definition, function, and interaction with importin α. *J. Biol. Chem.* **2007**, *282*, 5101–5105. [CrossRef] [PubMed]

54. Cautain, B.; Hill, R.; De Pedro, N.; Link, W. Components and regulation of nuclear transport processes. *FEBS J.* **2015**, *282*, 445–462. [CrossRef]

55. Kwan, J.A.; Schulze, C.J.; Wang, W.; Leon, H.; Sariahmetoglu, M.; Sung, M.; Sawicka, J.; Sims, D.E.; Sawicki, G.; Schulz, R. Matrix metalloproteinase-2 (MMP-2) is present in the nucleus of cardiac myocytes and is capable of cleaving poly (ADP-ribose) polymerase (PARP) in vitro. *FASEB J.* **2004**, *18*, 690–692. [CrossRef] [PubMed]

56. Si-Tayeb, K.; Monvoisin, A.; Mazzocco, C.; Lepreux, S.; Decossas, M.; Cubel, G.; Taras, D.; Blanc, J.F.; Robinson, D.R.; Rosenbaum, J. Matrix metalloproteinase 3 is present in the cell nucleus and is involved in apoptosis. *Am. J. Pathol.* **2006**, *169*, 1390–1401. [CrossRef] [PubMed]

57. Nakai, K.; Kanehisa, M. A knowledge base for predicting protein localization sites in eukaryotic cells. *Genomics* **1992**, *14*, 897–911. [CrossRef]

58. Eguchi, T.; Kubota, S.; Kawata, K.; Mukudai, Y.; Uehara, J.; Ohgawara, T.; Ibaragi, S.; Sasaki, A.; Kuboki, T.; Takigawa, M. Novel Transcription Factor-Like Function of Human Matrix Metalloproteinase 3 Regulating the CTGF/CCN2 Gene. *Mol. Cell. Biol.* **2008**, *28*, 2391–2413. [CrossRef]

59. Eguchi, T.; Calderwood, S.K.; Takigawa, M.; Kubota, S.; Kozaki, K. Intracellular MMP3 Promotes HSP Gene Expression in Collaboration with Chromobox Proteins. *J. Biogeogr.* **2017**, *118*, 43–51. [CrossRef]

60. Muromachi, K.; Kamio, N.; Narita, T.; Annen-Kamio, M.; Sugiya, H.; Matsushima, K. MMP-3 provokes CTGF/CCN2 production independently of protease activity and dependently on dynamin-related endocytosis, which contributes to human dental pulp cell migration. *J. Cell. Biochem.* **2012**, *113*, 1348–1358. [CrossRef]

61. Abdukhakimova, D.; Xie, Y. Comparative Analysis of NLS Sequence Suggests the Evolutionary Origin of Nuclear Matrix Metalloproteinase 7 during Cancer Evolution. *Int. J. Pharma Med. Biol. Sci.* **2016**, *5*, 206–210. [CrossRef]

62. Kiss, A.L.; Botos, E. Endocytosis via caveolae: Alternative pathway with distinct cellular compartments to avoid lysosomal degradation? *J. Cell. Mol. Med.* **2009**, *13*, 1228–1237. [CrossRef]

63. Lee, K.W.; Liu, B.; Ma, L.; Li, H.; Bang, P.; Koeffler, H.P.; Cohen, P. Cellular internalization of insulin-like growth factor binding protein-3. Distinct endocytic pathways facilitate re-uptake and nuclear localization. *J. Biol. Chem.* **2004**, *279*, 469–476. [CrossRef]

64. Gálvez, B.G.; Matías-Román, S.; Yáñez-Mó, M.; Vicente-Manzanares, M.; Sánchez-Madrid, F.; Arroyo, A.G. Caveolae Are a Novel Pathway for Membrane-Type 1 Matrix Metalloproteinase Traffic in Human Endothelial Cells. *Mol. Biol. Cell* **2004**, *15*, 678–687. [CrossRef]

65. Gasche, Y.; Copin, J.C.; Sugawara, T.; Fujimura, M.; Chan, P.H. Matrix metalloproteinase inhibition prevents oxidative stress-associated blood-brain barrier disruption after transient focal cerebral ischemia. *J. Cereb. Blood Flow Metab.* **2001**, *21*, 1393–1400. [CrossRef] [PubMed]

66. Si-Yayeb, K.; Monvoisin, A.; Mazzocco, C.; Lepreux, S.; Rosenbaum, J. Unexpected localization of the matrix metalloproteinase-3 (MMP-3) within the cell nucleus in liver cancer cells. Mechanisms and consequences. *J. Hepatol.* **2003**, *38*, 105. [CrossRef]

67. Ågren, M.S.; Schnabel, R.; Christensen, L.H.; Mirastschijski, U. Tumor necrosis factor-α-accelerated degradation of type I collagen in human skin is associated with elevated matrix metalloproteinase (MMP)-1 and MMP-3 ex vivo. *Eur. J. Cell Biol.* **2015**, *94*, 12–21. [CrossRef]

68. Xie, Y.; Lu, W.; Liu, S.; Yang, Q.; Shawn Goodwin, J.; Sathyanarayana, S.A.; Pratap, S.; Chen, Z. MMP7 interacts with ARF in nucleus to potentiate tumor microenvironments for prostate cancer progression in vivo. *Oncotarget* **2016**, *7*, 47609–47619. [CrossRef] [PubMed]

69. Zuo, X.; Pan, W.; Feng, T.; Shi, X.; Dai, J. Matrix metalloproteinase 3 promotes cellular anti-dengue virus response via interaction with transcription factor NFkB in cell nucleus. *PLoS ONE* **2014**, *9*, e84748. [CrossRef] [PubMed]

70. Marchant, D.J.; Bellac, C.L.; Moraes, T.J.; Wadsworth, S.J.; Dufour, A.; Butler, G.S.; Bilawchuk, L.M.; Hendry, R.G.; Robertson, A.G.; Cheung, C.T.; et al. A new transcriptional role for matrix metalloproteinase-12 in antiviral immunity. *Nat. Med.* **2014**, *20*, 493–502. [CrossRef] [PubMed]

71. Naphade, S.; Embusch, A.; Madushani, K.L.; Ring, K.L.; Ellerby, L.M. Altered Expression of Matrix Metalloproteinases and Their Endogenous Inhibitors in a Human Isogenic Stem Cell Model of Huntington's Disease. *Front. Neurosci.* **2018**, *11*, 736. [CrossRef]

72. Ayva, S.K.; Karabulut, A.A.; Akatli, A.N.; Atasoy, P.; Bozdogan, O. Epithelial expression of extracellular matrix metalloproteinase inducer/CD147 and matrix metalloproteinase-2 in neoplasms and precursor lesions derived from cutaneous squamous cells: An immunohistochemical study. *Pathol. Res. Pract.* **2013**, *209*, 627–634. [CrossRef]

73. Kivi, N.; Rönty, M.; Tarkkanen, J.; Auvinen, P.; Auvinen, E. Cell culture model predicts human disease: Altered expression of junction proteins and matrix metalloproteinases in cervical dysplasia. *BMC Clin. Pathol.* **2012**, *12*, 9. [CrossRef] [PubMed]

74. Shimizu-Hirota, R.; Xiong, W.; Baxter, B.T.; Kunkel, S.L.; Maillard, I.; Chen, X.W.; Sabeh, F.; Liu, R.; Li, X.Y.; Weiss, S.J. MT1-MMP regulates the PI3Kδ-Mi-2/NuRD-dependent control of macrophage immune function. *Genes Dev.* **2012**, *26*, 395–413. [CrossRef] [PubMed]

75. Limb, G.A.; Matter, K.; Murphy, G.; Cambrey, A.D.; Bishop, P.N.; Morris, G.E.; Khaw, P.T. Matrix metalloproteinase-1 associates with intracellular organelles and confers resistance to lamin A/C Degradation during apoptosis. *Am. J. Pathol.* **2005**, *166*, 1555–1563. [CrossRef]

76. Boström, P.; Söderström, M.; Vahlberg, T.; Söderström, K.O.; Roberts, P.J.; Carpén, O.; Hirsimäki, P. MMP-1 expression has an independent prognostic value in breast cancer. *BMC Cancer* **2011**, *11*, 348. [CrossRef]

77. Hazzaa, H.H.; Hager, E.A.A. Expression of MMP-1 and MMP-9 in localized aggressive periodontitis patients before and after treatment: A clinical and immunohistochemical study. *Egyptain Dent. J.* **2017**, *63*, 667–684. [CrossRef]

78. Malara, A.; Ligi, D.; Di Buduo, C.; Mannello, F.; Balduini, A. Sub-Cellular Localization of Metalloproteinases in Megakaryocytes. *Cells* **2018**, *7*, 80. [CrossRef]

79. Aldonyte, R.; Brantly, M.; Block, E.; Patel, J.; Zhang, J. Nuclear localization of active matrix metalloproteinase-2 in cigarette smoke-exposed apoptotic endothelial cells. *Exp. Lung Res.* **2009**, *35*, 59–75. [CrossRef]

80. Yang, Y.; Candelario-Jalil, E.; Thompson, J.F.; Cuadrado, E.; Estrada, E.Y.; Rosell, A.; Montaner, J.; Rosenberg, G.A. Increased intranuclear matrix metalloproteinase activity in neurons interferes with oxidative DNA repair in focal cerebral ischemia. *J. Neurochem.* **2010**, *112*, 134–149. [CrossRef]

81. Hill, J.W.; Poddar, R.; Thompson, J.F.; Rosenberg, G.A.; Yang, Y. Intranuclear matrix metalloproteinases promote DNA damage and apoptosis induced by oxygen-glucose deprivation in neurons. *Neuroscience* **2012**, *220*, 277–290. [CrossRef]

82. Yeghiazaryan, M.; Żybura-Broda, K.; Cabaj, A.; Włodarczyk, J.; Sławińska, U.; Rylski, M.; Wilczyński, G.M. Fine-structural distribution of MMP-2 and MMP-9 activities in the rat skeletal muscle upon training: A study by high-resolution in situ zymography. *Histochem. Cell Biol.* **2012**, *138*, 75–87. [CrossRef]

83. Hadler-Olsen, E.; Solli, A.I.; Hafstad, A.; Winberg, J.O.; Uhlin-Hansen, L. Intracellular MMP-2 activity in skeletal muscle is associated with type II fibers. *J. Cell. Physiol.* **2015**, *230*, 160–169. [CrossRef]

84. Aksenenko, M.B.; Kirichenko, A.K.; Ruksha, T.G. Russian study of morphological prognostic factors characterization in BRAF-mutant cutaneous melanoma. *Pathol. Res. Pract.* **2015**, *211*, 521–527. [CrossRef] [PubMed]

85. Aksenenko, M.B.; Ruksha, T.G. B oa ao aooa-2 aooppp6c-aooo aa ao o. *Russ. J. Skin Vener. Dis.* **2018**, *21*, 4–9. [CrossRef]

86. Kim, K.; Punj, V.; Kim, J.M.; Lee, S.; Ulmer, T.S.; Lu, W.; Rice, J.C.; An, W. MMP-9 facilitates selective proteolysis of the histone H3 tail at genes necessary for proficient osteoclastogenesis. *Genes Dev.* **2016**, *30*, 208–219. [CrossRef] [PubMed]

87. Machado, G.F.; Melo, G.D.; Moraes, O.C.; Souza, M.S.; Marcondes, M.; Perri, S.H.V.; Vasconcelos, R.O. Differential alterations in the activity of matrix metalloproteinases within the nervous tissue of dogs in distinct manifestations of visceral leishmaniasis. *Vet. Immunol. Immunopathol.* **2010**, *136*, 340–345. [CrossRef] [PubMed]

88. Tsai, J.P.; Liou, J.H.; Kao, W.T.; Wang, S.C.; Lian, J.D.; Chang, H.R. Increased Expression of Intranuclear Matrix Metalloproteinase 9 in Atrophic Renal Tubules Is Associated with Renal Fibrosis. *PLoS ONE* **2012**, *7*, e48164. [CrossRef]

89. Dandachi, N.G.; Shapiro, S.D. A protean protease: MMP-12 fights viruses as a protease and a transcription factor. *Nat. Med.* **2014**, *20*, 470–472. [CrossRef]

90. Mäkinen, L.K. Matrix Metalloproteinases and Toll-Like Receptors in Early-Stage Oral Tongue Squamous Cell Carcinoma. Ph.D. Thesis, University of Helsinki, Helsinki, Finland, 2015.

91. Hong, Y.; Kim, H.; Lee, S.; Jin, Y.; Choi, J.; Lee, S.R.; Chang, K.T.; Hong, Y. Role of melatonin combined with exercise as a switch-like regulator for circadian behavior in advanced osteoarthritic knee. *Oncotarget* **2017**, *8*, 97633–97647. [CrossRef]

92. Smith, B.; Bhowmick, N. Role of EMT in Metastasis and Therapy Resistance. *J. Clin. Med.* **2016**, *5*, 17. [CrossRef]

93. Scheau, C.; Badarau, I.A.; Costache, R.; Caruntu, C.; Mihai, G.L.; Didilescu, A.C.; Constantin, C.; Neagu, M. The role of matrix metalloproteinases in the epithelial-mesenchymal transition of hepatocellular carcinoma. *Anal. Cell. Pathol.* **2019**, *2019*, 9423907.

94. Sounni, N.E.; Roghi, C.; Chabottaux, V.; Janssen, M.; Munaut, C.; Maquoi, E.; Galvez, B.G.; Gilles, C.; Frankenne, F.; Murphy, G.; et al. Up-regulation of Vascular Endothelial Growth Factor-A by Active Membrane-type 1 Matrix Metalloproteinase Through Activation of Src-Tyrosine Kinases. *J. Biol. Chem.* **2004**, *279*, 13564–13574. [CrossRef] [PubMed]

95. Koppisetti, R.K.; Fulcher, Y.G.; Jurkevich, A.; Prior, S.H.; Xu, J.; Overduin, M.; Van Doren, S.R. Ambidextrous Binding of Cell and Membrane Bilayers by Soluble Matrix Metalloproteinase-12. *Nat. Commun.* **2014**, *5*, 5552. [CrossRef] [PubMed]

96. Freudenberg, J.A.; Chen, W.T. Induction of Smad1 by MT1-MMP contributes to tumor growth. *Int. J. Cancer* **2007**, *121*, 966–977. [CrossRef] [PubMed]

97. Ali, M.A.M.; Garcia-Vilas, J.A.; Cromwell, C.R.; Hubbard, B.P.; Hendzel, M.J.; Schulz, R. Matrix metalloproteinase-2 mediates ribosomal RNA transcription by cleaving nucleolar histones. *bioRxiv* **2020**. [CrossRef]

98. Saeb-Parsy, K.; Veerakumarasivam, A.; Wallard, M.J.; Thorne, N.; Kawano, Y.; Murphy, G.; Neal, D.E.; Mills, I.G.; Kelly, J.D. MT1-MMP regulates urothelial cell invasion via transcriptional regulation of Dickkopf-3. *Br. J. Cancer* **2008**, *99*, 663–669. [CrossRef]

99. Nagasawa, T.; Arai, M.; Togari, A. Inhibitory effect of minocycline on osteoclastogenesis in mouse bone marrow cells. *Arch. Oral Biol.* **2011**, *56*, 924–931. [CrossRef]

100. Kim, Y.; Kim, J.; Lee, H.; Shin, W.R.; Lee, S.; Lee, J.; Park, J.I.; Jhun, B.H.; Kim, Y.H.; Yi, S.J.; et al. Tetracycline analogs inhibit osteoclast differentiation by suppressing MMP-9-Mediated Histone H3 cleavage. *Int. J. Mol. Sci.* **2019**, *20*, 4038. [CrossRef]

101. Kimura-Ohba, S.; Yang, Y. Oxidative DNA Damage Mediated by Intranuclear MMP Activity Is Associated with Neuronal Apoptosis in Ischemic Stroke. *Oxid. Med. Cell. Longev.* **2016**, *2016*, 6927328. [CrossRef]

102. Pirici, D.; Pirici, I.; Mogoanta, L.; Margaritescu, O.; Tudorica, V.; Margaritescu, C.; Ion, D.A.; Simionescu, C.; Coconu, M. Matrix metalloproteinase-9 expression in the nuclear compartment of neurons and glial cells in aging and stroke. *Neuropathology* **2012**, *32*, 492–504. [CrossRef]

103. Galluzzi, L.; Vitale, I.; Aaronson, S.A.; Abrams, J.M.; Adam, D.; Agostinis, P.; Alnemri, E.S.; Altucci, L.; Amelio, I.; Andrews, D.W.; et al. Molecular mechanisms of cell death: Recommendations of the Nomenclature Committee on Cell Death 2018. *Cell Death Differ.* **2018**, *25*, 486–541. [CrossRef]

104. Wang, X.; Lee, S.R.; Arai, K.; Lee, S.R.; Tsuji, K.; Rebeck, G.W.; Lo, E.H. Lipoprotein receptor-mediated induction of matrix metalloproteinase by tissue plasminogen activator. *Nat. Med.* **2003**, *9*, 1313–1317. [CrossRef] [PubMed]

105. Bergers, G.; Brekken, R.; McMahon, G.; Vu, T.H.; Itoh, T.; Tamaki, K.; Tanzawa, K.; Thorpe, P.; Itohara, S.; Werb, Z.; et al. Matrix metalloproteinase-9 triggers the angiogenic switch during carcinogenesis. *Nat. Cell Biol.* **2000**, *2*, 737–744. [CrossRef] [PubMed]

106. Pfeffer, C.M.; Singh, A.T.K. Apoptosis: A target for anticancer therapy. *Int. J. Mol. Sci.* **2018**, *19*, 448. [CrossRef] [PubMed]

107. Heutinck, K.M.; ten Berge, I.J.M.; Hack, C.E.; Hamann, J.; Rowshani, A.T. Serine proteases of the human immune system in health and disease. *Mol. Immunol.* **2010**, *47*, 1943–1955. [CrossRef]

108. Nissinen, L.; Kähäri, V.M. Matrix metalloproteinases in inflammation. *Biochim. Biophys. Acta Gen. Subj.* **2014**, *1840*, 2571–2580. [CrossRef]

109. Soria-Valles, C.; Gutiérrez-Fernández, A.; Osorio, F.G.; Carrero, D.; Ferrando, A.A.; Colado, E.; Fernández-García, M.S.; Bonzon-Kulichenko, E.; Vázquez, J.; Fueyo, A.; et al. MMP-25 Metalloprotease Regulates Innate Immune Response through NF-κB Signaling. *J. Immunol.* **2016**, *197*, 296–302. [CrossRef]

110. Elkington, P.T.G.; O'Kane, C.M.; Friedland, J.S. The paradox of matrix metalloproteinases in infectious disease. *Clin. Exp. Immunol.* **2005**, *142*, 12–20. [CrossRef]

111. Chen, Y.L.; Li, W.Y.; Hu, J.J.; Li, Y.; Liu, G.M.; Jin, T.C.; Cao, M.J. Nucleus-translocated matrix metalloprotease 1 regulates innate immune response in Pacific abalone (Haliotis discus hannai). *Fish Shellfish Immunol.* **2019**, *84*, 290–298. [CrossRef]

112. Gross, J.; Lapiere, C.M. Collagenolytic activity in amphibian tissues: A tissue culture assay. *Proc. Natl. Acad. Sci. USA* **1962**, *48*, 1014–1022. [CrossRef]

113. Liotta, L.A.; Thorgeirsson, U.P.; Garbisa, S. Role of collagenases in tumor cell invasion. *Cancer Metastasis Rev.* **1982**, *1*, 277–288. [CrossRef]

114. Amour, A.; Knight, C.G.; Webster, A.; Slocombe, P.M.; Stephens, P.E.; Knäuper, V.; Docherty, A.J.P.; Murphy, G. The in vitro activity of ADAM-10 is inhibited by TIMP-1 and TIMP-3. *FEBS Lett.* **2000**, *473*, 275–279. [CrossRef]

115. Winer, A.; Adams, S.; Mignatti, P. Matrix metalloproteinase inhibitors in cancer therapy: Turning past failures into future successes. *Mol. Cancer Ther.* **2018**, *17*, 1147–1155. [CrossRef]

116. Kwok, H.F.; Botkjaer, K.A.; Tape, C.J.; Huang, Y.; McCafferty, J.; Murphy, G. Development of a "mouse and human cross-reactive" affinity-matured exosite inhibitory human antibody specific to TACE (ADAM17) for cancer immunotherapy. *Protein Eng. Des. Sel.* **2014**, *27*, 179–190. [CrossRef] [PubMed]

117. Huang, Y.; Benaich, N.; Tape, C.; Kwok, H.F.; Murphy, G. Targeting the sheddase activity of Adam17 by an Anti-ADAM17 antibody D1(A12) inhibits head and neck squamous cell carcinoma cell proliferation and motility via blockage of bradykinin induced HERs transactivation. *Int. J. Biol. Sci.* **2014**, *10*, 702–714. [CrossRef] [PubMed]

118. Botkjaer, K.A.; Kwok, H.F.; Terp, M.G.; Karatt-Vellatt, A.; Santamaria, S.; McCafferty, J.; Andreasen, P.A.; Itoh, Y.; Ditzel, H.J.; Murphy, G. Development of a specific affinity-matured exosite inhibitor to MT1-MMP that efficiently inhibits tumor cell invasion in vitro and metastasis in vivo. *Oncotarget* **2016**, *7*, 16773–16792. [CrossRef]

119. Fields, G.B. New strategies for targeting matrix metalloproteinases. *Matrix Biol.* **2015**, *44–46*, 239–246. [CrossRef]

120. Tammam, S.N.; Azzazy, H.M.E.; Lamprecht, A. How successful is nuclear targeting by nanocarriers? *J. Control. Release* **2016**, *229*, 140–153. [CrossRef]

121. Pan, L.; He, Q.; Liu, J.; Chen, Y.; Ma, M.; Zhang, L.; Shi, J. Nuclear-targeted drug delivery of tat peptide-conjugated monodisperse mesoporous silica nanoparticles. *J. Am. Chem. Soc.* **2012**, *134*, 5722–5725. [CrossRef]

122. Austin, L.A.; Kang, B.; Yen, C.W.; El-Sayed, M.A. Plasmonic imaging of human oral cancer cell communities during programmed cell death by nuclear-targeting silver nanoparticles. *J. Am. Chem. Soc.* **2011**, *133*, 17594–17597. [CrossRef]

123. Austin, L.A.; Kang, B.; Yen, C.W.; El-Sayed, M.A. Nuclear targeted silver nanospheres perturb the cancer cell cycle differently than those of nanogold. *Bioconjug. Chem.* **2011**, *22*, 2324–2331. [CrossRef]

124. Fan, W.; Shen, B.; Bu, W.; Zheng, X.; He, Q.; Cui, Z.; Zhao, K.; Zhang, S.; Shi, J. Design of an intelligent sub-50 nm nuclear-targeting nanotheranostic system for imaging guided intranuclear radiosensitization. *Chem. Sci.* **2015**, *6*, 1747–1753. [CrossRef] [PubMed]

125. Webb, A.H.; Gao, B.T.; Goldsmith, Z.K.; Irvine, A.S.; Saleh, N.; Lee, R.P.; Lendermon, J.B.; Bheemreddy, R.; Zhang, Q.; Brennan, R.C.; et al. Inhibition of MMP-2 and MMP-9 decreases cellular migration, and angiogenesis in in vitro models of retinoblastoma. *BMC Cancer* **2017**, *17*, 434. [CrossRef] [PubMed]

126. Gialeli, C.; Theocharis, A.D.; Karamanos, N.K. Roles of matrix metalloproteinases in cancer progression and their pharmacological targeting. *FEBS J.* **2011**, *278*, 16–27. [CrossRef] [PubMed]
127. Wojtowicz-Praga, S.M.; Dickson, R.B.; Hawkins, M.J. Matrix metalloproteinase inhibitors. *Investig. New Drugs* **1997**, *15*, 61–75. [CrossRef] [PubMed]
128. Cathcart, J.; Pulkoski-Gross, A.; Cao, J. Targeting matrix metalloproteinases in cancer: Bringing new life to old ideas. *Genes Dis.* **2015**, *2*, 26–34. [CrossRef] [PubMed]

Publisher's Note: MDPI stays neutral with regard to jurisdictional claims in published maps and institutional affiliations.

 *biology*

Review

Matrix Metalloproteases in Pancreatic Ductal Adenocarcinoma: Key Drivers of Disease Progression?

Etienne J. Slapak [1,2,3], JanWillem Duitman [1,2], Cansu Tekin [1,2,3], Maarten F. Bijlsma [2,3] and C. Arnold Spek [1,2,*]

1 Center of Experimental and Molecular Medicine, University of Amsterdam, Amsterdam UMC, 1105 AZ Amsterdam, The Netherlands; e.j.slapak@amsterdamumc.nl (E.J.S.); j.w.duitman@amsterdamumc.nl (J.D.); c.tekin@amsterdamumc.nl (C.T.)
2 Laboratory for Experimental Oncology and Radiobiology, Cancer Center Amsterdam, University of Amsterdam, Amsterdam UMC, 1105 AZ Amsterdam, The Netherlands; m.f.bijlsma@amsterdamumc.nl
3 Oncode Institute, 1105 AZ Amsterdam, The Netherlands
* Correspondence: c.a.spek@amsterdamumc.nl

Received: 26 March 2020; Accepted: 15 April 2020; Published: 18 April 2020

Abstract: Pancreatic cancer is a dismal disorder that is histologically characterized by a dense fibrotic stroma around the tumor cells. As the extracellular matrix comprises the bulk of the stroma, matrix degrading proteases may play an important role in pancreatic cancer. It has been suggested that matrix metalloproteases are key drivers of both tumor growth and metastasis during pancreatic cancer progression. Based upon this notion, changes in matrix metalloprotease expression levels are often considered surrogate markers for pancreatic cancer progression and/or treatment response. Indeed, reduced matrix metalloprotease levels upon treatment (either pharmacological or due to genetic ablation) are considered as proof of the anti-tumorigenic potential of the mediator under study. In the current review, we aim to establish whether matrix metalloproteases indeed drive pancreatic cancer progression and whether decreased matrix metalloprotease levels in experimental settings are therefore indicative of treatment response. After a systematic review of the studies focusing on matrix metalloproteases in pancreatic cancer, we conclude that the available literature is not as convincing as expected and that, although individual matrix metalloproteases may contribute to pancreatic cancer growth and metastasis, this does not support the generalized notion that matrix metalloproteases drive pancreatic ductal adenocarcinoma progression.

Keywords: MMP; MMP2; MMP9; MMP7; MMP14; matrix metalloproteases; PDAC; pancreatic cancer

1. Introduction

Pancreatic ductal adenocarcinoma (PDAC) is a devastating disease with the worst survival outcome of any cancer [1]. Its incidence, which is around 10 per 100,000 individuals, is rising in developed countries [2,3], with 458 thousand new cases and 432 thousand deaths in 2018 worldwide [4]. The 5-year survival rate is around 9%, and the 10-year mortality is approaching 99% [5]. Progress towards improving survival has been slow, and current treatment options are inadequate. The only significant progress that has been made is in the form of lower mortality rates for patients eligible for resections, and a slight prolongation and improved quality of life in patients with inoperable disease with the use of chemotherapeutic agents. Single-agent gemcitabine treatment has been the standard of care for inoperable PDAC for many years, although the observed benefits are small in daily practice [6–9] and seem restricted to patients with a good performance status [10]. More recently, nanoparticle albumin-bound paclitaxel was shown to exert superior antitumor activity compared to gemcitabine monotherapy, thereby establishing nab-paclitaxel and gemcitabine combination therapy as first-line chemotherapy regimens in PDAC [11]. In patients with a good performance status, combination therapy with folinic acid, fluorouracil, irinotecan and oxaliplatin (FOLFIRINOX) is superior over other treatments [12] and FOLFIRINOX is consequently emerging as the new standard of care for relatively fit patients [13]. Importantly however, even in the specific group of patients eligible for FOLFIRINOX treatment, the survival benefit is limited [14].

1.1. Tumor Microenvironment of PDAC

PDAC is characterized by a strong desmoplastic reaction, which results in an archetypal tumor microenvironment, consisting of a dense stroma surrounding the tumor cells [15,16]. The stroma forms the bulk of the tumor, taking up to 90% of the total tumor mass and consists of many cellular and acellular components like (myo)fibroblasts, macrophages, blood vessels and extracellular matrix components such as, among others, collagen I, collagen IV, laminin and fibronectin. In the stroma, the extracellular matrix has traditionally been considered to be a stable structure that mainly plays a supportive role in maintaining tissue morphology. Nowadays, however, it is evident that the extracellular matrix forms a dynamic and versatile milieu that affects the fundamental processes of the surrounding cells [17,18]. Accordingly, the loss of extracellular matrix homeostasis and integrity is considered one of the hallmarks of cancer and typically defines transitional events, resulting in cancer progression and metastasis [19]. Moreover, the loss of extracellular matrix homeostasis due to stromal depletion aggravates pancreatic cancer progression in preclinical animal models [20–22].

1.2. Matrix Metalloproteases in the Tumor Microenvironment

The desmoplastic PDAC stroma contains many different proteases that play a key role in the crosstalk between tumor and stromal cells. An intriguing group of proteases in the tumor microenvironment consist of matrix metalloproteases (MMPs), which are primarily known for their ability to degrade extracellular matrix components. Altered expression and/or activity of MMPs in the tumor microenvironment is likely to lead to the loss of homeostasis of the extracellular matrix, thereby driving PDAC progression. Based upon this notion, MMPs are considered important contributors to PDAC progression and experimental PDAC studies frequently use MMPs as surrogate markers for treatment responses. Decreased MMP levels are, nowadays, considered as important signs of the anti-tumorigenic potential of the gene/compound/miRNA under study. In the current review, we address whether the literature supports the concept that MMPs drive PDAC progression and if decreased MMP levels under experimental settings are indicative of the treatment response. To this end, we performed a systematic review of patient and experimental animal studies, focusing on MMPs in PDAC.

1.3. Overview of Matrix Metalloproteases

MMPs are calcium-dependent zinc-containing endopeptidases of the metzincin protease superfamily. They typically contain an N-terminal propeptide of approximately 80–90 amino acids, with a conserved PRCGXPD motif that is responsible for maintaining latency via the binding of the cysteine residue to the zinc atom in the active site [23]. After the proteolytic removal of the propeptide, the active form of MMP contains a calcium-dependent catalytic domain of around 200 amino acids, which contains a hydrophobic S1'-pocket that determines substrate specificity, proceeded by a linker region of variable length, and the C-terminal hemopexin-like domain, which spans approximately 200 amino acids. The hemopexin-like domain, which is absent in some MMP family members, plays a functional role in substrate binding and/or in interactions with tissue inhibitors of metalloproteases (TIMPs), a family of specific MMP protein inhibitors [24].

Since the identification of a diffusible collagenolytic factor in living amphibian tissue that is capable of degrading undenatured calf skin collagen [25], a total of 24 MMPs have been identified in humans [26]. According to their substrate specificity, MMPs are classified into subfamilies: (1) collagenases, (2) gelatinases, (3) stromelysins, (4) matrilysins, (5) membrane-type MMPs and

(6) others. Despite the general acceptance of the classification system based on extracellular matrix substrates, MMPs are rather promiscuous in substrate recognition and also proteolytically cleave substrates beyond extracellular matrix proteins.

2. Methods

To provide a comprehensive overview of the role of MMPs in PDAC, a systematic PubMed search without restrictions was performed. A combination of the search terms "pancreatic cancer" and every individual MMP (both using the official gene name and the common name; see Supplementary Materials Table S1) was used to retrieve papers published up to 1 March 2020. All papers were independently screened by their title and abstract, followed by full text assessment to include papers that contained MMP expression analysis in PDAC patients and papers that contained animal experiments that targeted (either genetically or pharmacologically) MMPs in pancreatic cancer models. The excluded papers were those that contained in vitro data only, papers that assayed MMP levels in experimental animal models without interventions or genetic modifications, or papers that did not focus on PDAC.

3. Results

We retrieved 64 papers focusing on collagenases, 642 papers focusing on gelatinases, 51 papers focusing on stromelysins, 93 papers focusing on matrilysins, 66 papers focusing on transmembrane MMPs and 21 papers focusing on other MMPs (Figure 1). After the removal of duplicates, 816 eligible studies were identified and were vigorously screened to obtain those that contained patient data and/or animal experiments in which MMPs were targeted. This resulted in the inclusion of 14 papers focusing on collagenases, 60 on gelatinases, 11 on stromelysins, 21 on matrilysins, 12 on transmembrane-type MMPs and five on the so-called "other" MMPs. As several of the eligible papers contained data on multiple MMPs, the total number of papers including patient/experimental animal data selected for the review was 91.

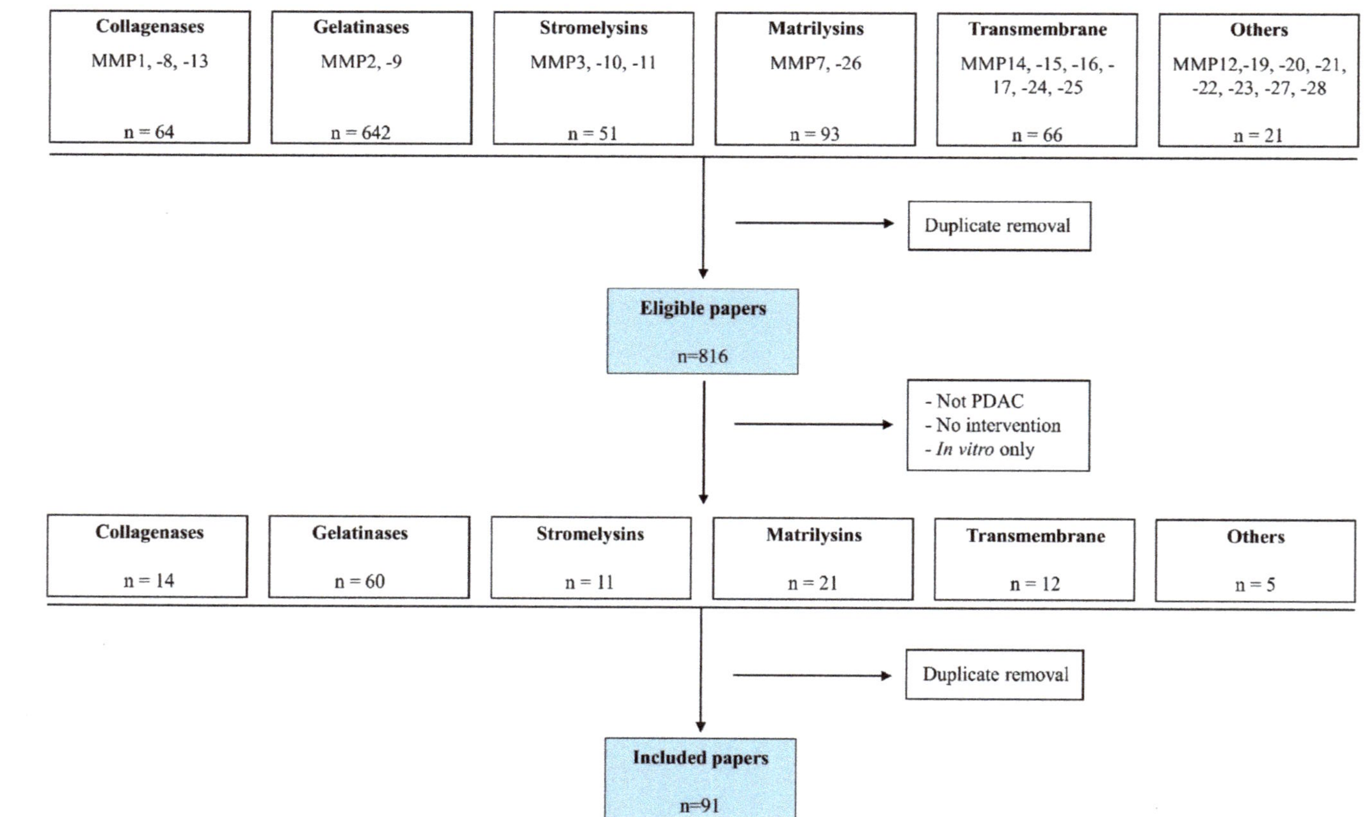

Figure 1. Flowchart of paper inclusion. Using the search criteria indicated in Supplementary Materials Table S1, we obtained 814 eligible papers that we screened for the presence of patient and/or matrix metalloprotease (MMP) intervention in animal models. After the exclusion of duplicate papers, we ended up with 91 papers that were included in the review.

3.1. Collagenases in PDAC

Despite the general notion that collagenases (MMP1, MMP8 and MMP13) are key players in cancer biology [27–29], relatively little is known about collagenases in PDAC. Although MMP-1 is consistently shown to be overexpressed in PDAC patients compared to healthy controls [30–36], its effects on cancer progression are inconsistent (Table 1). For example, MMP1 overexpression has been reported as being associated with both a poor prognosis [30] and prolonged survival [37], although no correlations with tumor size, differentiation status and lymph node involvement have been observed [30,36,38]. Despite an elegant recent study showing that MMP1-dependent protease activated receptor (PAR)-1 drives PDAC cell migration and perineural invasion [33], the important role of MMP1 in PDAC is not supported by the experimental data. Besides MMP1 overexpression, MMP8 [36,39] and MMP13 [34,40] are also overexpressed in PDAC patients compared to healthy controls. The relevance of increased MMP expression is not well documented and only a single study showed that MMP-13 expression is associated with lymph node metastasis and the tumor's pathological stage [41]. Interestingly however, MMP13 overexpression significantly promoted the invasion of the PDAC cells in vitro, whereas MMP13 inhibition blocked leptin-mediated PDAC cell invasion [41], while CD40 agonist-dependent resolution of fibrosis and enhanced chemotherapy efficacy were diminished by MMP13 inhibition [42].

3.2. Gelatinases in PDAC

The most studied MMPs in PDAC are, without a doubt, the gelatinases (MMP2 and MMP9; see Figure 1). The vast majority of studies show that both MMP2 [34,35,43–62] and MMP9 [34,36,39,48,49,53,54,59,62–67] are upregulated in PDAC patients (Table 1), while a minority of studies fail to show a difference in expression between PDAC and the controls [36,38,52,60,61,68–71]. The potential clinical relevance is less pronounced, as just half of the studies reported associations between increased MMP2 or MMP9 levels with clinical characteristics such as survival, metastasis or tumor stage [43,46–48,50,51,53,56–58,61,63,65,67,68,72,73], whereas in the other half of the studies no such correlations were observed (Table 2). Despite the rather diverse observations in patients, initial preclinical experimental animal experiments showed promising results (Table 3). Batimastat treatment of mice harboring orthotopic pancreatic cancers reduced cancer growth, metastasis and death compared to control-treated mice, while also potentiating gemcitabine sensitivity [74–77]. Batimastat was also shown to reduce metastasis and death when PDAC cells were directly injected into the spleen of recipient mice, in order to mimic liver metastasis in PDAC [78]. Although batimastat is not specific to MMP2 and MMP9 and also inhibits MMP1, MMP3, MMP7, MMP8 and several ADAM family members, based on the gelatin zymography of tumor samples before and after treatment, it was hypothesized that the tumor-inhibiting effect of batimastat was dependent on MMP2 and, to a lesser extent, MMP9. The potential importance of MMP2 and MMP9 in PDAC progression is further supported by studies using more specific inhibitors like MMI-166, RO28-2653 and OPB-3206. Indeed, the selective MMP2, MMP9 and MMP14 inhibitor MMI-166 inhibited PDAC growth in both mice and Syrian hamsters [79,80], whereas RO28-2653 and OPB-3206 (both also selective MMP2, MMP9 and MMP14 inhibitors) reduced chemically induced pancreatic carcinogenesis in Syrian hamsters [81,82]. Finally, treatment with the selective MMP2 and MMP9 inhibitor SB-3CT reduced the lung metastasis of subcutaneously implanted PDAC cells [83].

The most conclusive evidence of the role of MMP2 in PDAC progression comes from subcutaneous models, in which the injection of shMMP2-silenced PANC1 cells resulted in smaller tumors compared to the injection of control shRNA transduced cells [84], whereas treatment with MMP2-blocking peptides limited tumor growth and angiogenesis [85].

In a similar way to the inconclusive association studies in patients (see above and Table 2), experimental animal experiments specifically targeting MMP9 show inconsistent results (Table 3). Orthotopic injections of MMP9-overexpressing Panc02 cells led to bigger tumors than injections of their control counterparts, but the absence/presence of MMP9 did not affect metastasis [86]. Treatment with a MMP9-blocking antibody did not affect the tumor growth of subcutaneously implanted PDAC cells, but did enhance gemcitabine and nab-paclitaxel sensitivity when PDAC cells were injected into the peritoneal cavity [87]. Doxycycline treatment, suggested to specifically target MMP9, reduced the growth of subcutaneously injected Capan-1 cells [88]. Finally, subcutaneous or orthotopic implantation of PDAC cells in MMP9-deficient mice diminished tumor take, tumor growth, angiogenesis and metastasis [83,89] but tumor progression and metastasis increased in MMP9-deficient mice on the *Kras(G12D)/Tp53* background [90].

3.3. Stromelysins in PDAC

Clinical studies do not support the general role of stromelysins (MMP3, MMP10 and MMP11) in PDAC (Table 1). Although MMP11 is consistently upregulated and associated with clinical characteristics in PDAC patients [35,36,91–93], the data for MMP3 is more controversial. Only half of the studies focusing on MMP3 suggest its expression is increased in PDAC patients compared to control tissue [34,35,94,95], and only a single study suggests that MMP3 is associated with patient survival [95]. Besides clinical studies, preclinical animal models also do not support an important role for stromelysins in PDAC progression. Apart from a study which suggests, but does not prove, that MMP10 drives the invasion and metastasis of PDAC [96], it has only been shown that MMP3 overexpression on the *Kras(G12D)* background increases neoplastic alterations in pancreatic acinar cells [94]. These premalignant morphological changes were accompanied by the recruitment of infiltrating immune cells and the expression of smooth muscle actin and collagen, indicating that MMP3 is not only a coconspirator of Kras in inducing tumorigenic changes in epithelial cells, but also that it promotes the establishment of a tumorigenic microenvironment. Though it has been suggested that MMP3 may play a role in PDAC initiation, the actual importance of endogenous MMP3 (as opposed to overexpressed MMP3) in PDAC progression and its potential clinical relevance remains elusive.

3.4. Matrilysins in PDAC

MMP7 and MMP26 are the only two members of the matrilysin subfamily. A large number of studies have compared MMP7 expression in PDAC patients with pancreatitis patients and/or healthy controls and have consistently shown that MMP7 levels are elevated in PDAC patients (Table 1) [34–36,54,69,91,97–104]. More importantly, MMP7 levels correlate with metastasis and/or survival in most, but not all, studies. Based upon these reports, it is suggested that MMP7 is an important regulator of tumor formation. In line with this notion, preclinical experimental animal models show that MMP7 expression is intimately linked with acinar-to-ductal metaplasia and that pancreatic duct ligation-dependent acinar cell loss, caspase-3 activation, and subsequent metaplasia is significantly reduced in MMP7-deficient mice (Table 3) [98]. The effect of MMP7 on acinar-to-ductal metaplasia seems model-specific, however, as MMP7 deficiency did not affect pancreatitis driven-PanIN development in Pfta1-Cre Kras(G12D) mice [105]. In addition to PDAC initiation, MMP7 also seems to drive PDAC progression. Using several genetic Kras-driven PDAC models, it was shown that both tumor size and metastasis were significantly reduced by MMP7 deficiency. The percentage of mice with lymph node metastasis reduced from around 60 in MMP7-proficient mice to 0 in MMP7-deficient mice, whereas the percentage of mice with liver metastasis dropped from 67% to 13% due to MMP7 deficiency [105]. In line with these findings, the metastasis of MMP7-silenced PANC1 cells was largely reduced compared to control PANC1 cells, whereas pharmacological MMP7 inhibition with sulfur-2-(4-chlorine-3-trifluoromethyl phenyl)-sulfonamido-4-phenylbutyric acid (SCTPSPA) also significantly reduced the metastasis of PANC1 cells [101]. MMP26 expression was also induced in PDAC patients compared to the controls and, intriguingly, MMP26 was expressed significantly more often in tumors with lymph node involvement. Although this is suggestive of the general role of matrilysins in PDAC progression, experimental data confirming the pro-tumorigenic role of MMP26 in PDAC is lacking and it remains to be established whether MMP26 is indeed a driver of disease progression or merely acts as a marker of PDAC metastasis [106].

3.5. Membrane-Type MMPs in PDAC

Seven membrane-bound MMPs have been described so far: the transmembrane members MMP14, MMP15, MMP16, MMP23 and MMP24, and the GPI-anchored members MMP17 and MMP25. Of the membrane-bound MMPs, MMP14 seems most relevant in the setting of PDAC (Tables 1–3). Indeed, the overexpression of MMP14 in mice expressing an activating Kras(G12D) mutation led to more large, dysplastic mucin-containing papillary lesions compared to the control Kras(G12D) mice (Table 3) [107]. Using subcutaneous models, MMP14 overexpression in cancer cells seems to reduce the cytotoxic effect of gemcitabine [108], whereas MMP14 inhibition in pancreatic stellate cells limits tumor growth [84]. Moreover, the cancer cell-specific overexpression of membrane-type 1 matrix metalloproteinase cytoplasmic tail binding protein-1 (MTCBP-1; MMP14 binding protein inhibiting its activity) restricts metastasis in orthotopic PDAC models, further suggesting that MMP14 may enhance tumor progression [109]. However, clinical data do not support the important role of MMP14 in PDAC progression (Tables 1 and 2). Although MMP14 may be overexpressed in PDAC [44,110], MMP14 does not correlate with clinical characteristics such as tumor differentiation, tumor size, lymph node status, or patient survival [31,37,111].

3.6. Other MMPs in PDAC

The so-called other MMPs (i.e., MMP12, MMP19, MMP20, MMP21, MMP27 and MMP28) are not very well characterized in PDAC. Although some members seem to be overexpressed in PDAC [106,111,112] and may be associated with tumor stage and patients survival (Table 1) [111–113], no preclinical studies have addressed the role of these MMPs in PDAC (Table 2). Therefore, their actual importance remains to be established.

3.7. Clinical Trials with MMP Inhibitors in PDAC

Only two phase 3 trials focusing on MMP inhibition in PDAC have been published [114,115]. One trial showed that the addition of marimastat (a broad-spectrum MMP inhibitor targeting MMP1, MMP2, MMP7, MMP9 and MMP14) to gemcitabine in a double-blind placebo-controlled, randomized study was well-tolerated but did not show clinical benefits in PDAC patients [114]. The overall response rates (11% and 16% with and without the addition of marimastat, respectively), progression-free survival and time to treatment failure were similar in both treatment arms. Another phase 3 trial showed that BAY 12-9566 (tanomastat; MMP2, MMP3 and MMP9 inhibitor) treatment was also well tolerated by PDAC patients but was inferior to gemcitabine, with median survival times of 3.74 and 6.59 months for the BAY 12-9566 and gemcitabine arm, respectively [115]. Median progression-free survival and quality-of-life analyses also favored gemcitabine, arguing against MMP inhibition in the setting of PDAC.

The fact that there are no clinical benefits obtained through MMP inhibition does not imply that MMPs do not contribute to PDAC progression. As elegantly discussed [116,117], the disappointing clinical trial results may be due to several reasons, of which the inclusion of advanced stage disease seems most relevant. Broad spectrum MMP inhibitors may also lack efficacy as they could block the potential tumor inhibitory activities of specific MMPs. As indicated above, MMP9 deficiency on the *Kras(G12D)* background enhanced tumor progression and invasive growth [90], supporting this notion and providing an alternative explanation for the negative marimastat and BAY 12-9566 results in PDAC patients. Finally, the poor clinical efficacy of MMP inhibitors could also be explained by the overestimation of the role of MMPs in PDAC progression based on preclinical models that do not fully capture the complexity of human disease.

Table 1. MMP expression levels in Pancreatic ductal adenocarcinoma (PDAC) patients and controls. Red indicates increased MMP levels, blue indicates no difference and green indicates decreased MMP levels in PDAC patients.

Member	Patient Number	Method	Difference	Reference
MMP1	45 PC, 10 CO	IHC	no difference	[36]
MMP1	8 PC, 8 CO	RNA	no difference	[59]
MMP1	248 PC, 216 CO	Serum	no difference	[69]
MMP1	46 PC, 5 CO	IHC	up vs healthy	[30]
MMP1	25 PC	RNA	up vs adjacent CO	[31]
MMP1	10 PC, 12 CP, 5 CO	IHC	up vs CO	[32]
MMP1	45 PC	RNA	up vs adjacent CO	[33]
MMP1	30 PC	IHC	up vs adjacent CO	[33]
MMP1	104 PC, 62 CO	IHC	up vs CO	[34]
MMP1	17 PC, 17 CO	RNA	up vs CO	[35]
MMP1	18 PC, 8 CO	RNA	up vs healthy	[36]
MMP2	75 PC, 10 CO	IHC	no difference	[36]
MMP2	18 PC, 8 CO	RNA	no difference	[36]
MMP2	70 PC and 10 CO	IHC	no difference	[38]
MMP2	92 PC, 43 CP, 91 CO	Serum	no difference	[68]
MMP2	35 PC	RNA/IHC	no difference	[70]
MMP2	46 PC, 13 CO	Serum	no difference	[71]
MMP2	104 PC, 62 CO	IHC	up vs CO	[34]
MMP2	17 PC, 17 CO	RNA	upvsCO	[35]
MMP2	122 PC	IHC	up vs adjacent CO	[43]
MMP2	18 PC, 9 CP, 9 CO	RNA	up vs both others	[44]
MMP2	12 PC, 11 CP, 7 CO	pancreatic juice	up vs both others	[45]
MMP2	127 PC	IHC	up vs CO	[46]
MMP2	20 PC	IHC	up vs CO	[47]
MMP2	32 PC, 31 CP	ELISA on tissue	up vs CP	[48]
MMP2	110 PC, 24 BT	Plasma	up vs BT	[49]
MMP2	37 PC, 7 CP	IHC	up vs CP and CO	[50]
MMP2	45 PC	IHC	up vs CO	[51]
MMP2	51 PC, 60 CO	Urine	up vs CO	[52]
MMP2	44 PC, 8 CO	IHC	up vs CO	[52]
MMP2	30 PC, 17 CO	IHC	up vs CO	[53]
MMP2	29 PC	IHC	up vs adjacent CO	[54]
MMP2	127 PC, 25 CP, 25 CO	Plasma	up vs CP and CO	[55]
MMP2	106 PC	RNA/WB	up vs adjacent CO	[56]
MMP2	40 PC, 10 CO	IHC	up vs CO	[57]
MMP2	67 PC, 20 CO	IHC	up vs adjacent CO	[58]
MMP2	8 PC, 8 CO	RNA	upvsCO	[59]
MMP2	10 PC, 3 CO	ZG	upvsCO	[60]
MMP2	33 PC, 14 CP, 13 CO	ZG/WB	upvsCO	[61]
MMP2	22 PC, 9 CP, 9 CO	RNA	up vs adjacent CO	[62]
MMP2	10 PC, 213 CO	Serum	down vs CO	[118]
MMP3	45 PC, 10 CO	IHC	no difference	[36]
MMP3	18 PC, 8 CO	RNA	no difference	[36]
MMP3	8 PC, 8 CO	RNA	no difference	[59]
MMP3	104 PC, 62 CO	IHC	up vs CO	[34]
MMP3	17 PC, 17 CO	RNA	up vs CO	[35]
MMP3	140 PC, 12 CO	IHC	up vs CO	[94]
MMP3	140 PC, 12 CO	IHC	up vs CO	[95]
MMP7	18 PC, 8 CO	RNA	no difference	[36]
MMP7	104 PC, 62 CO	IHC	up vs CO	[34]
MMP7	17 PC, 17 CO	RNA	up vs CO	[35]
MMP7	45 PC, 10 CO	IHC	up vs CO	[36]
MMP7	29 PC	IHC	up vs adjacent CO	[54]
MMP7	248 PC, 216 CO	Serum	up vs CO	[68]

Table 1. *Cont.*

Member	Patient Number	Method	Difference	Reference
MMP7	44 PC, 17 CP	RNA	up vs CP	[91]
MMP7	70 PC	RNA	up vs adjacent CO	[97]
MMP7	32 PC, ? CO	IHC	up vs CO	[98]
MMP7	47 PC, 10 CO	IHC	up vs CO	[99]
MMP7	63 PC, 31 CP	Plasma	up vs CP	[100]
MMP7	30 PC	RNA	up vs adjacent CO	[101]
MMP7	5 PC, 5 CP, 62 CO	IHC	up vs CP and CO	[102]
MMP7	131 PC, 30 CP, 131 CO	Plasma	up vs CO	[103]
MMP7	10 PC	RNA	up vs adjacent CO	[104]
MMP8	248 PC, 216 CO	Serum	no difference	[69]
MMP8	75 PC, 10 CO	IHC	up vs CO	[36]
MMP8	91 PC, 41 CP, 30 CO	RNA (PBMCs)	up vs CO	[39]
MMP9	18 PC, 8 CO	RNA	no difference	[36]
MMP9	70 PC, 10 CO	IHC	no difference	[38]
MMP9	51 PC, 60 CO	urine	no difference	[52]
MMP9	10 PC, 3 CO	ZG	no difference	[60]
MMP9	33 PC, 14 CP, 13 CO	ZG/WB	no difference	[61]
MMP9	248 PC, 216 CO	Serum	no difference	[69]
MMP9	35 PC	RNA/IHC	no difference	[70]
MMP9	104 PC, 62 CO	IHC	up vs CO	[34]
MMP9	45 PC, 10 CO	IHC	up vs CO	[36]
MMP9	91 PC, 41 CP, 30 CO	RNA (PBMCs)	up vs CP and CO	[39]
MMP9	32 PC, 31 CP	ELISA on tissue	up vs CP	[48]
MMP9	110 PC, 24 BT	Plasma	up vs BT	[49]
MMP9	30 PC, 17 CO	IHC	up vs CO	[53]
MMP9	29 PC	IHC	up vs adjacent CO	[54]
MMP9	8 PC, 8 CO	RNA	up vs CO	[59]
MMP9	22 PC, 9 CP, 9 CO	RNA	up vs adjacent CO	[62]
MMP9	36 PC	IHC	up vs CO	[63]
MMP9	9 PC, 9 CO	MS/MS	up vs CO	[64]
MMP9	78 PC, 45 CP, 70 CO	Serum	up vs both	[65]
MMP9	62 PC, 16 CO	IHC	up vs CO	[66]
MMP9	103 PC, 6 CO	IHC	up vs CO	[67]
MMP10	17 PC, 17 CO	RNA	no difference	[35]
MMP11	17 PC, 17 CO	RNA	up vs CO	[35]
MMP11	18 PC, 8 CO	RNA	up vs CO	[36]
MMP11	75 PC, 10 CO	IHC	up vs CO	[36]
MMP11	44 PC, 17 CP	RNA	up vs CP	[91]
MMP11	12 PC, 16 CO	Blood	up vs CO	[92]
MMP11	21 PC, 9 CO	IHC	up vs CO	[93]
MMP12	75 PC, 10 CO	IHC	no difference	[36]
MMP12	39 PC, 13 CO	RNA/WB/IHC	up vs CO	[111]
MMP13	104 PC, 62 CO	IHC	up vs CO	[34]
MMP13	45 PC	RNA	up vs adjacent CO	[40]
MMP14	75 PC, 10 CO	IHC	no difference	[36]
MMP14	35 PC	RNA/IHC	no difference	[111]
MMP14	18 PC, 9 CP, 9 CO	RNA	up vs both others	[44]
MMP14	64 PC, 9 CO	IHC	up vs CO	[110]
MMP15	18 PC, 9 CP, 9 CO	RNA	up vs both others	[44]
MMP15	18 PC, 8 CO	RNA	reduced vs CO	[36]
MMP16	18 PC, 9 CP, 9 CO	RNA	no difference	[44]
MMP16	12 PC	IHC	up vs adjacent CO	[119]
MMP19	102 PC	IHC	up vs adjacent CO	[112]
MMP20	102 PC	IHC	up vs adjacent CO	[112]
MMP21	25 PC, 18 CO	IHC	up vs CO	[106]
MMP26	25 PC, 18 CO	IHC	up vs CO	[106]

Pancreatic cancer (PC); pancreatitis (CP); healthy control (CO); benign tumor (BT); immunohistochemistry (IHC); Western blot (WB); zymography (DG).

Table 2. Association between MMP expression and clinical characteristics of PDAC. Red indicates that MMP levels are associated with poor outcome, blue indicates no association and green indicates that MMP levels are associated with improved survival.

Member	Patient Number	Method	Correlation		Reference
MMP1	45 PC, 10 CO	IHC		no	[36]
MMP1	70 PC	IHC		no	[38]
MMP1	46 PC, 5 CO	IHC	OS,	LM, Size, Stage	[30]
MMP1	30 PC	IHC		PNI	[33]
MMP1	51 PC	IHC/serum		OS	[37]
MMP2	75 PC, 10 CO	IHC		no	[36]
MMP2	70 PC, 10 CO	IHC		no	[38]
MMP2	51 PC	IHC/serum		no	[37]
MMP2	32 PC, 31 CP	ELISA on tissue		no	[48]
MMP2	37 PC, 7 CP	IHC		no	[50]
MMP2	29 PC	IHC		no	[54]
MMP2	127 PC, 25 CP, 25 CO	plasma		no	[55]
MMP2	35 PC	RNA/IHC		no	[70]
MMP2	32 PC	IHC		no	[120]
MMP2	67 PC	IHC	LM,	PNI, OS, DF	[121]
MMP2	122 PC	IHC		OS, DF	[43]
MMP2	127 PC	IHC		OS, Stage	[46]
MMP2	20 PC	IHC		LM	[47]
MMP2	37 PC, 7 CP	IHC		LM, DM	[50]
MMP2	45 PC	IHC		OS, LM, Stage	[51]
MMP2	30 PC, 17 CO	IHC		LM, Stage, Size	[53]
MMP2	106 PC	RNA/WB		DM, Stage	[56]
MMP2	40 PC, 10 CO	IHC		LM	[57]
MMP2	67 PC, 20 CO	IHC		LM, Stage, PNI	[58]
MMP2	33 PC, 14 CP, 13 CO	ZG/WB		Stage	[61]
MMP2	92 PC, 43 CP, 91 CO	serum		LM, DM	[68]
MMP2	32 PC	IHC		VI	[72]
MMP2	88 PC	IHC		OS	[73]
MMP3	45 PC, 10 CO	IHC		no	[36]
MMP3	18 PC, 8 CO	RNA		no	[36]
MMP3	70 PC	IHC		no	[38]
MMP3	140 PC, 12 CO	IHC		OS	[95]
MMP7	51 PC	IHC/serum		no	[37]
MMP7	29 PC	IHC		no	[54]
MMP7	88 PC	IHC		no	[73]
MMP7	45 PC, 10 CO	IHC	OS,	LM, DIF, Stage	[36]
MMP7	70 PC	IHC	OS,	Size, DIF	[38]
MMP7	134 PC	IHC	Stage, PNI,	OS	[122]
MMP7	70 PC	RNA		LM, Size	[97]
MMP7	47 PC, 10 CO	IHC		OS, DM	[99]
MMP7	10 PC	RNA		OS	[104]
MMP7	101 PC	serum		OS	[105]
MMP7	39 PC	IHC		LM, OS	[123]
MMP8	75 PC, 10 CO	IHC		no	[36]
MMP8	91 PC, 41 CP, 30 CO	RNA (PBMCs)		no	[39]
MMP9	45 PC, 10 CO	IHC		no	[36]
MMP9	70 PC, 10 CO	IHC		no	[38]
MMP9	51 PC	IHC/serum		no	[37]
MMP9	91 PC, 41 CP, 30 CO	RNA (PBMCs)		no	[39]
MMP9	29 PC	IHC		no	[54]
MMP9	33 PC, 14 CP, 13 CO	ZG/WB		no	[61]
MMP9	9 PC, 9 CO	MS/MS		no	[64]
MMP9	35 PC	RNA/IHC		no	[70]
MMP9	32 PC	IHC		no	[123]
MMP9	27 PC	IHC		no	[124]
MMP9	62 PC, 16 CO	IHC	PNI,	LM, Stage, Size	[66]
MMP9	63 PC	IHC	VI,	OS, LM, DM	[125]
MMP9	62 PC	IHC	LM,	OS	[126]
MMP9	32 PC, 31 CP	ELISA on tissue		LM	[48]
MMP9	30 PC, 17 CO	IHC		LM, Stage, Size	[53]
MMP9	36 PC	IHC		LM, DM	[63]
MMP9	78 PC, 45 CP, 70 CO	serum		OS	[65]

Table 2. *Cont.*

Member	Patient Number	Method	Correlation		Reference
MMP9	103 PC, 6 CO	IHC	OS, LM, DM, VI, Stage		[67]
MMP9	32 PC	IHC	VI		[72]
MMP9	88 PC	IHC	OS, DF,	DM	[73]
MMP10	51 PC	IHC/serum	no		[37]
MMP11	75 PC, 10 CO	IHC	OS, LM,	DIF, Size	[36]
MMP11	not indicated	RNA	OS		[92]
MMP12	75 PC, 10 CO	IHC	no		[36]
MMP12	39 PC, 13 CO	RNA/WB/IHC	OS,	LM, Stage	[111]
MMP13	60 PC	IHC	LM		[41]
MMP14	70 PC	IHC	no		[38]
MMP14	75 PC, 10 CO	IHC	no		[36]
MMP14	37 PC	RNA/IHC	no		[70]
MMP15	78 PC	IHC	OS, DF,	PNI, LM, DM, Stage	[127]
MMP19	102 PC	IHC	OS, DF, PNI, Stage		[112]
MMP20	102 PC	IHC	OS, DF, Stage, PNI		[112]
MMP21	25 PC, 18 CO	IHC	no		[106]
MMP26	25 PC, 18 CO	IHC	LM		[106]
MMP28	not indicated	RNA	OS		[113]

Pancreatic cancer (PC); pancreatitis (CP); healthy control (CO); benign tumor (BT); immunohistochemistry (IHC); Western blot (WB); zymography (DG); overall survival (OS); disease-free survival (DF); lymph node metastasis (LM); perineural invasion (PNI); venous invasion (VI); distant metastasis (DM); differentiation (DIF).

Table 3. Experimental animal models that target MMPs.

Target	Model	"Treatment"	Result	Reference
MMP1	Sciatic nerve invasion	shMMP1 PANC1 cells	Reduced perineural invasion	[33]
MMP2/9?	Orthotopic injection HPAC cells	Batimastat (day −7 till death/sacrifice)	Increased gemcitabine sensitivity, No effect single treatment	[74]
	Orthotopic injection HPAC cells	Batimastat (day −4 till death/sacrifice)	Reduced tumor growth, metastasis and death	[75]
	Orthotopic injection HPAC cells	Batimastat (day 7 till death/sacrifice)	Reduced local invasion and death	[76]
	Orthotopic injection HPAC cells	Batimastat (day 7 till death/sacrifice)	Reduced tumor weight	[77]
	Injection AsPC1 or Capan-1 cells in spleen	Batimastat (day −7 till day 14)	Reduced metastasis and death	[78]
	Subcutaneous injection SW1990 cells	MMI-166 from day 7 till sacrifice at day 28	Reduced tumor growth	[79]
	Orthotopic injection PGHAM cells (Syrian hamster)	MMI-166 (day 1 till sacrifice)	Reduced tumor growth, liver metastasis and MVD	[80]
	BOP injections (Syrian hamster)	RO28-2653 (week 6 till week 14)	Reduced liver metastasis, No effect death	[81]
	BOP injections (Syrian hamster)	OPB-3206 in diet from day 48 till sacrifice	Reduced invasive carcinoma	[82]
	Subcutaneous injection Panc02 or MIAPaca2 cells	SB-3CT (day 1 till sacrifice)	Reduced lung metastasis	[83]
MMP2	Subcutaneous injection organoid and PSC	shMMP2 PSC	Reduced tumor growth	[84]
	Subcutaneous injection PANC-1 or CFPAC-1 cells	MMP2 blocking peptides after tumor take	Reduced growth and MVD	[85]
MMP3	Kras(G12D) mice	MMP3 overexpression	Increased neoplastic alterations	[94]
MMP7	Ductal ligation	MMP7 deficient mice	Reduced ductal metaplasia	[98]
	Pfta1-Cre/KrasG12D mice	MMP7 deficiency	No effect acinar to ductal metaplasia	[105]
	Pdx1-Crelate/KrasG12D mice	MMP7 deficiency	Reduced tumor development	[105]
	Pdx1-CreLate/KrasG12D/p53f/+ mice	MMP7 deficiency	Reduced tumor growth and metastasis	[105]
	Tail vein injection PANC1 cells	shMMP7	Reduced liver and lung metastasis	[101]
		SCTPSPA (day −2 till day 25)	Reduced lung metastasis	[101]
MMP9	Subcutaneous injection Panc02 cells	MMP9 deficient mice	Reduced lung metastasis	[83]
	Orthotopic injection Panc02 cells	MMP9 overexpression	Enhanced tumor growth, No effect metastasis	[86]
	Subcutaneous injection AsPC-1 cells	aMMP9 antibody AB0046 (day 1 till day 14)	No effect on tumor weight	[87]
	Injection AsPC-1 cells in peritoneal cavity	aMMP9 antibody AB0046 (day 14 till day 56)	Increased gemcitabine/nab-paclitaxel sensitivity, No effect metastasis	[87]
	Subcutaneous injection Capan-1 cells	Doxycycline (day 1 till day 14)	Reduced growth and MVD	[88]
	Orthotopic injection L3.6pl cells	MMP9 deficient mice	Reduced tumor take, growth and MVD	[89]
	Pdx-1+/Cre;KrasG12D;Trp53 mice	MMP9 deficiency	Increased progression and invasive growth	[90]
	Intravenous injection 9801 or Panc02 cells	MMP9 deficient mice	Increased metastasis	[90]
MMP14	Subcutaneous injection organoid and PSC	shMMP14 PSC	Reduced tumor growth	[84]
	KrasG12D mice	MMP14 overexpression	Increased number of PanIN lesions	[107]
	Subcutaneous injection PANC1 or HPAF-II cells	MMP14 overexpression	Reduced gemcitabine sensitivity, No effect single treatment	[108]
	Orthotopic injection DanG or BxPc3 cells	MTCBP-1 overexpression	Reduced metastasis, No effect tumor growth	[109]

Note: All experiments were performed using mice unless indicated otherwise.

4. Conclusions

The potential clinical relevance of MMPs in PDAC has largely been addressed using patient-derived tumor material. These studies show a rather consistent picture with respect to MMP overexpression in tumors compared to control sections, although almost 25% of the studies do not show significant differences between patients and controls. However, the association of MMP overexpression with clinical characteristics is not as convincing as suggested in the literature. Half of the studies show that high MMP levels are associated with (lymph node) metastasis and reduced survival, whereas the other half of the studies do not show any correlation with clinical characteristics. Patient-derived data do not, therefore, seem to allow firm conclusions that MMP expression levels (in general) are associated with PDAC progression and poor prognosis to be drawn, especially when considering that publication bias may have resulted in negative studies not being published.

Initial preclinical experimental animal models using broad spectrum MMP inhibitors are more in line with the general role of MMPs in PDAC progression, as different inhibitors limit tumor growth and metastasis in subcutaneous, orthotopic and spontaneous PDAC models. The contribution of individual MMPs in PDAC progression is, however, not very well established. Only MMP2, MMP7 and MMP14 are shown to potentiate tumor growth and/or metastasis in multiple independent papers. For others, the literature is conflicting or missing and no clear conclusions can be drawn. Importantly, however, conflicting results do not indicate that the individual MMPs have no effect in PDAC. The biology of PDAC and MMP is complex and MMPs may act in a context-dependent manner, with both tumor-promoting and tumor-inhibiting effects. The conflicting role of MMP9 serves as an excellent example for this notion. The data rather convincingly show that tumor MMP9 expression drives PDAC progression, but systemic MMP9 ablation triggers invasive growth and metastasis by blocking MMP9-dependent tumor-inhibiting effects in the bone marrow.

Despite the presence of a large range of MMP-deficient animals and the relative ease of generating MMP deficient cells with CRISPR technology, the majority of MMPs have not been studied in preclinical PDAC animal models. To fully appreciate the importance of individual MMPs in PDAC progression and to assess their potential clinical relevance, we have to await studies that combine (pharmacological inhibition in) genetic Kras-driven spontaneous models with subcutaneous and/or orthotopic models, in which MMPs are specifically depleted in stromal or tumor cells. In particular, experiments that address pharmacological treatment with specific MMP inhibitors after tumors could turn out to be invaluable for establishing the context-dependent role of individual MMPs in PDAC. Before such studies have been performed, we should be careful not to generalize the available literature.

Although broad spectrum MMP inhibitors limit PDAC progression in preclinical animal models [73–82], they seem to lack efficacy in a clinical setting [115,116]. This disparity between preclinical data and clinical trials can be attributed to several factors—for instance, differences in pharmacokinetics, pharmacodynamics and metabolism and the failure to accurately model the tumor microenvironment [128]. In particular, xenograft models, which lack a functional immune system, show a reduced complexity and cellular diversity compared to human disease models. Moreover, the degree of aneuploidy in human tumors results in great variety within inter-tumoral gene modifications, in a different manner compared to how it occurs in mice [129,130]. All of these species-related differences limit the capacity of preclinical mouse models to accurately predict the response of MMP inhibitors in PDAC patients.

In conclusion, based on our systematic review on the role of matrix metalloproteases in PDAC, we conclude that the available literature is not as consistent as envisioned and that, although individual matrix metalloproteases seem to contribute to PDAC growth and metastasis, our review does not support the generalized notion that matrix metalloproteases drive PDAC progression.

Supplementary Materials: The following are available online at http://www.mdpi.com/2079-7737/9/4/80/s1, Table S1: Search terms used and number of papers retrieved.

Funding: This research was funded by grants from the Dutch Cancer Foundation (UVA 2017-11174 and UVA 2014-6782) and the Netherlands Organization for Scientific Research (VENI grant 016.186.046).

Conflicts of Interest: The authors declare no conflict of interest. M.F.B. has acted as a consultant to Servier, and received research funding from Celgene.

References

1. Neoptolemos, J.P.; Urrutia, R.; Abbruzzese, J.L.; Büchler, M.W. (Eds.) *Pancreatic Cancer*; Springer: New York, NY, USA, 2018; Volume XXXIII, pp. 3–18.
2. Rawla, P.; Sunkara, T.; Gaduputi, V. Epidemiology of pancreatic cancer: Global Trends, Etiology and Risk Factors. *World J. Oncol.* **2019**, *10*, 10–27. [CrossRef]
3. Simard, E.P.; Ward, E.M.; Siegel, R.; Jemal, A. Cancers with increasing incidence trends in the United States: 1999 through 2008. *CA Cancer J. Clin.* **2012**, *62*, 118–128. [CrossRef]
4. Bray, F.; Ferlay, J.; Soerjomataram, I.; Siegel, R.L.; Torre, L.A.; Jemal, A. Global cancer statistics 2018: GLOBOCAN estimates of incidence and mortality worldwide for 36 cancers in 185 countries. *CA Cancer J. Clin.* **2018**, *68*, 394–424. [CrossRef]
5. Siegel, R.L.; Miller, K.D.; Jemal, A. Cancer statistics 2019. *CA Cancer J. Clin.* **2019**, *69*, 7–34. [CrossRef]
6. Burris, H.A., III; Moore, M.J.; Andersen, J.; Green, M.R.; Rothenberg, M.L.; Modiano, M.R.; Cripps, M.C.; Portenoy, R.K.; Storniolo, A.M.; Tarassoff, P.; et al. Improvements in survival and clinical benefit with gemcitabine as first-line therapy for patients with advanced pancreas cancer: A randomized trial. *J. Clin. Oncol.* **1997**, *15*, 2403–2413. [CrossRef]
7. Cunningham, D.; Chau, I.; Stocken, D.D.; Valle, J.W.; Smith, D.; Steward, W.; Harper, P.G.; Dunn, J.; Tudur-Smith, C. Phase III randomized comparison of gemcitabine versus gemcitabine plus capecitabine in patients with advanced pancreatic cancer. *J. Clin. Oncol.* **2009**, *27*, 5513–5518. [CrossRef]
8. Heinemann, V.; Quietzsch, D.; Gieseler, F.; Gonnermann, M.; Schönekäs, H.; Rost, A.; Neuhaus, H.; Haag, C.; Clemens, M.; Heinrich, B.; et al. Randomized phase III trial of gemcitabine plus cisplatin compared with gemcitabine alone in advanced pancreatic cancer. *J. Clin. Oncol.* **2006**, *24*, 3946–3952. [CrossRef]
9. Louvet, C.; Labianca, R.; Hammel, P.; Lledo, G.; Zampino, M.G.; André, T.; Zaniboni, A.; Ducreux, M.; Aitini, E.; Taïeb, J.; et al. Gemcitabine in combination with oxaliplatin compared with gemcitabine alone in locally advanced or metastatic pancreatic cancer: Results of a GERCOR and GISCAD phase III trial. *J. Clin. Oncol.* **2005**, *23*, 3509–3516. [CrossRef]
10. Reni, M.; Cordio, S.; Milandri, C.; Passoni, P.; Bonetto, E.; Oliani, C.; Luppi, G.; Nicoletti, R.; Galli, L.; Bordonaro, R.; et al. Gemcitabine versus cisplatin, epirubicin, fluorouracil, and gemcitabine in advanced pancreatic cancer: A randomised controlled multicentre phase III trial. *Lancet Oncol.* **2005**, *6*, 369–376. [CrossRef]
11. Awasthi, N.; Zhang, C.; Schwarz, A.M.; Hinz, S.; Wang, C.; Williams, N.S.; Schwarz, M.A.; Schwarz, R.E. Comparative benefits of Nab-paclitaxel over gemcitabine or polysorbate-based docetaxel in experimental pancreatic cancer. *Carcinogenesis* **2013**, *34*, 2361–2369. [CrossRef]
12. Heinemann, V.; Haas, M.; Boeck, S. Systemic treatment of advanced pancreatic cancer. *Cancer Treat. Rev.* **2012**, *38*, 843–853. [CrossRef]
13. Rombouts, S.J.; Mungroop, T.H.; Heilmann, M.N.; van Laarhoven, H.W.; Busch, O.R.; Molenaar, I.Q.; Besselink, M.G.; Wilmink, J.W. FOLFIRINOX in Locally Advanced and Metastatic Pancreatic Cancer: A Single Centre Cohort Study. *J. Cancer* **2016**, *7*, 1861–1866. [CrossRef]
14. Suker, M.; Beumer, B.R.; Sadot, E.; Marthey, L.; Faris, J.E.; Mellon, E.A.; El-Rayes, B.F.; Wang-Gillam, A.; Lacy, J.; Hosein, P.J.; et al. FOLFIRINOX for locally advanced pancreatic cancer: A systematic review and patient level meta-analysis. *Lancet Oncol.* **2016**, *17*, 801–810. [CrossRef]
15. Hidalgo, M. Pancreatic cancer. *N. Engl. J. Med.* **2010**, *362*, 1605–1617. [CrossRef]
16. Bijlsma, M.; van Laarhoven, H. The conflicting roles of tumor stroma in pancreatic cancer and their contribution to the failure of clinical trials: A systematic review and critical appraisal. *Cancer Metastasis Rev.* **2015**, *34*, 97–114. [CrossRef]
17. Hynes, R.O. The extracellular matrix: Not just pretty fibrils. *Science* **2009**, *326*, 1216–1219. [CrossRef]
18. Lu, P.; Weaver, V.M.; Werb, Z. The extracellular matrix: A dynamic niche in cancer progression. *J. Cell Biol.* **2012**, *196*, 395–406. [CrossRef]

19. Filipe, E.C.; Chitty, J.L.; Cox, T.R. Charting the unexplored extracellular matrix in cancer. *Int. J. Exp. Pathol.* **2018**, *99*, 58–76. [CrossRef]

20. Özdemir, B.C.; Pentcheva-Hoang, T.; Carstens, J.L.; Zheng, X.; Wu, C.; Simpson, T.R.; Laklai, H.; Sugimoto, H.; Kahlert, C.; Novitskiy, S.V.; et al. Depletion of carcinoma-associated fibroblasts and fibrosis induces immunosuppression and accelerates pancreas cancer with reduced survival. *Cancer Cell* **2014**, *25*, 719–734. [CrossRef]

21. Rhim, A.D.; Oberstein, P.E.; Thomas, D.H.; Mirek, E.T.; Palermo, C.F.; Sastra, S.A.; Dekleva, E.N.; Saunders, T.; Becerra, C.P.; Tattersall, I.W.; et al. Stromal elements act to restrain, rather than support, pancreatic ductal adenocarcinoma. *Cancer Cell* **2014**, *25*, 735–747. [CrossRef]

22. Catenacci, D.V.; Junttila, M.R.; Karrison, T.; Bahary, N.; Horiba, M.N.; Nattam, S.R.; Marsh, R.; Wallace, J.; Kozloff, M.; Rajdevet, L.; et al. Randomized Phase Ib/II Study of Gemcitabine Plus Placebo or Vismodegib, a Hedgehog Pathway Inhibitor, in Patients With Metastatic pancreatic cancer. *J. Clin. Oncol.* **2015**, *33*, 4284–4292. [CrossRef]

23. Massova, I.; Kotra, L.P.; Fridman, R.; Mobashery, S. Matrix metalloproteinases: Structures, evolution, and diversification. *FASEB J.* **1998**, *12*, 1075–1095. [CrossRef]

24. Brew, K.; Dinakarpandian, D.; Nagase, H. Tissue inhibitors of metalloproteinases: Evolution, structure and function. *Biochim. Biophys. Acta* **2000**, *1477*, 267–283. [CrossRef]

25. Gross, J.; Lapiere, C. Collagenolytic activity in amphibian tissues: A tissue culture assay. *Proc. Natl. Acad. Sci. USA* **1962**, *48*, 1014–1022. [CrossRef]

26. Mittal, R.; Patel, A.P.; Debs, L.H.; Nguyen, D.; Patel, K.; Grati, M.; Mittal, J.; Yan, D.; Chapagain, P.; Liu, X.Z. Intricate Functions of Matrix Metalloproteinases in Physiological and Pathological Conditions. *J. Cell Physiol.* **2016**, *231*, 2599–2621. [CrossRef]

27. Juurikka, K.; Butler, G.S.; Salo, T.; Nyberg, P.; Åström, P. The Role of MMP8 in Cancer: A Systematic Review. *Int. J. Mol. Sci.* **2019**, *20*, 4506. [CrossRef]

28. Foley, C.J.; Kuliopulos, A. Mouse Matrix metalloprotease-1a (Mmp1a) Gives New Insight into MMP Function. *J. Cell Physiol.* **2014**, *229*, 1875–1880. [CrossRef]

29. Huang, S.H.; Law, C.H.; Kuo, P.H.; Hu, R.Y.; Yang, C.C.; Chung, T.W.; Li, J.M.; Lin, L.H.; Liu, Y.C.; Liao, E.C.; et al. MMP13 is involved in oral cancer cell metastasis. *Oncotarget* **2016**, *7*, 17144–171161.

30. Ito, T.; Ito, M.; Shiozawa, J.; Naito, S.; Kanematsu, T.; Sekine, I. Expression of the MMP-1 in Human Pancreatic Carcinoma: Relationship with Prognostic Factor. *Mod. Pathol.* **1999**, *12*, 669–674.

31. Rogers, A.; Smith, M.J.; Doolan, P.; Clarke, C.; Clynes, M.; Murphy, J.F.; McDermott, A.; Swan, N.; Crotty, P.; Ridgway, P.F.; et al. Invasive Markers Identified by Gene Expression Profiling in pancreatic cancer. *Pancreatology* **2012**, *12*, 130–140. [CrossRef]

32. Tahara, H.; Sato, K.; Yamazaki, Y.; Ohyama, T.; Horiguchi, N.; Hashizume, H.; Kakizaki, S.; Takagi, H.; Ozaki, I.; Arai, H.; et al. Transforming Growth Factor-α Activates Pancreatic Stellate Cells and May Be Involved in Matrix metalloproteinase-1 Upregulation. *Lab. Investig.* **2013**, *93*, 720–732. [CrossRef]

33. Huang, C.; Li, Y.; Guo, Y.; Guo, Y.; Zhang, Z.; Lian, G.; Chen, Y.; Li, J.; Su, Y.; Li, J.; et al. MMP1/PAR1/SP/NK1R paracrine loop modulates early perineural invasion of pancreatic cancer cells. *Theranostics* **2018**, *8*, 3074–3086. [CrossRef]

34. Ohlund, D.; Ardnor, B.; Oman, M.; Naredi, P.; Sund, M. Expression pattern and circulating levels of endostatin in patients with pancreas cancer. *Int. J. Cancer* **2008**, *122*, 2805–2810. [CrossRef]

35. Bramhall, S.R.; Neoptolemos, J.P.; Stamp, G.W.; Lemoine, N.R. Imbalance of expression of matrix metalloproteinases (MMPs) and tissue inhibitors of the matrix metalloproteinases (TIMPs) in human pancreatic carcinoma. *J. Pathol.* **1997**, *182*, 347–355. [CrossRef]

36. Jones, L.E.; Humphreys, M.J.; Campbell, F.; Neoptolemos, J.P.; Boyd, M.T. Comprehensive analysis of matrix metalloproteinase and tissue inhibitor expression in pancreatic cancer: Increased expression of matrix metalloproteinase-7 predicts poor survival. *Clin. Cancer Res.* **2004**, *10*, 2832–2845. [CrossRef]

37. Kahlert, C.; Fiala, M.; Musso, G.; Halama, N.; Keim, S.; Mazzone, M.; Lasitschka, F.; Pecqueux, M.; Klupp, F.; Schmidt, T.; et al. Prognostic Impact of a Compartment-Specific Angiogenic Marker Profile in Patients With pancreatic cancer. *Oncotarget* **2014**, *5*, 12978–12989. [CrossRef]

38. Yamamoto, H.; Itoh, F.; Iku, S.; Adachi, Y.; Fukushima, H.; Sasaki, S.; Mukaiya, M.; Hirata, K.; Imai, K. Expression of Matrix Metalloproteinases and Tissue Inhibitors of Metalloproteinases in Human Pancreatic Adenocarcinomas: Clinicopathologic and Prognostic Significance of Matrilysin Expression. *J. Clin. Oncol.* **2001**, *19*, 1118–1127. [CrossRef]

39. Moz, S.; Basso, D.; Padoan, A.; Padoan, A.; Bozzato, D.; Fogar, P.; Zambon, C.F.; Pelloso, M.; Sperti, C.; Vigili de Kreutzenberg, S.; et al. Blood expression of matrix metalloproteinases 8 and 9 and of their inducers S100A8 and S100A9 supports diagnosis and prognosis of pancreatic cancer -associated diabetes mellitus. *Clin. Chim. Acta* **2016**, *456*, 24–30. [CrossRef]

40. Tréhoux, S.; Duchêne, B.; Jonckheere, N.; Van Seuningen, I. The MUC1 oncomucin regulates pancreatic cancer cell biological properties and chemoresistance. Implication of p42–44 MAPK, Akt, Bcl-2 and MMP13 pathways. *Biochem. Biophys. Res. Commun.* **2015**, *456*, 757–762. [CrossRef]

41. Fa, Y.; Gan, Y.; Shen, Y.; Cai, X.; Song, Y.; Zhao, F.; Yao, M.; Gu, J.; Tu, H. Leptin signaling enhances cell invasion and promotes the metastasis of human pancreatic cancer via increasing MMP-13 production. *Oncotarget* **2015**, *6*, 16120–16134.

42. Long, K.B.; Gladney, W.L.; Tooker, G.M.; Graham, K.; Fraietta, J.A.; Beatty, G.L. IFNγ and CCL2 Cooperate to Redirect Tumor-Infiltrating Monocytes to Degrade Fibrosis and Enhance Chemotherapy Efficacy in Pancreatic Carcinoma. *Cancer Discov.* **2016**, *6*, 400–413. [CrossRef] [PubMed]

43. Zhai, L.L.; Cai, C.Y.; Wu, Y.; Tang, Z.G. Correlation and prognostic significance of MMP-2 and TFPI-2 differential expression in pancreatic carcinoma. *Int. J. Clin. Exp. Pathol.* **2015**, *8*, 682–691.

44. Ellenrieder, V.; Alber, B.; Lacher, U.; Hendler, S.F.; Menke, A.; Boeck, W.; Wagner, M.; Wilda, M.; Friess, H.; Büchler, M.; et al. Role of MT-MMPs and MMP-2 in pancreatic cancer progression. *Int. J. Cancer* **2000**, *85*, 14–20. [CrossRef]

45. Yokoyama, M.; Ochi, K.; Ichimura, M.; Mizushima, T.; Shinji, T.; Koide, N.; Tsurumi, T.; Hasuoka, H.; Harada, M. Matrix metalloproteinase-2 in pancreatic juice for diagnosis of pancreatic cancer. *Pancreas* **2002**, *24*, 344–347. [CrossRef]

46. Juuti, A.; Lundin, J.; Nordling, S.; Louhimo, J.; Haglund, C. Epithelial MMP-2 expression correlates with worse prognosis in pancreatic cancer. *Oncology* **2006**, *71*, 61–68. [CrossRef]

47. He, Y.; Liu, X.D.; Chen, Z.Y.; Zhu, J.; Xiong, Y.; Li, K.; Dong, J.H.; Li, X. Interaction between cancer cells and stromal fibroblasts is required for activation of the uPAR-uPA-MMP-2 cascade in pancreatic cancer metastasis. *Clin. Cancer Res.* **2007**, *13*, 3115–3124. [CrossRef]

48. Lekstan, A.; Lampe, P.; Lewin-Kowalik, J.; Olakowski, M.; Jablonska, B.; Labuzek, K.; Jedrzejowska-Szypulka, H.; Olakowska, E.; Gorka, D.; Filip, I.; et al. Concentrations and activities of metalloproteinases 2 and 9 and their inhibitors (TIMPS) in chronic pancreatitis and pancreatic adenocarcinoma. *J. Physiol. Pharmacol.* **2012**, *63*, 589–599.

49. Śmigielski, J.; Piskorz, Ł.; Talar-Wojnarowska, R.; Malecka-Panas, E.; Jabłoński, S.; Brocki, M. The estimation of metaloproteinases and their inhibitors blood levels in patients with pancreatic tumors. *World J. Surg. Oncol.* **2013**, *11*, 137. [CrossRef]

50. Singh, N.; Das, P.; Datta Gupta, S.; Sahni, P.; Pandey, R.M.; Gupta, S.; Chauhan, S.S.; Saraya, A. Prognostic significance of extracellular matrix degrading enzymes-cathepsin L and matrix metalloproteases-2 [MMP-2] in human pancreatic cancer. *Cancer Investig.* **2013**, *31*, 461–471. [CrossRef]

51. Xu, L.; Ding, X.; Tan, H.; Qian, J. Correlation between B7-H3 expression and matrix metalloproteinases 2 expression in pancreatic cancer. *Cancer Cell Int.* **2013**, *13*, 81. [CrossRef] [PubMed]

52. Roy, R.; Zurakowski, D.; Wischhusen, J.; Frauenhoffer, C.; Hooshmand, S.; Kulke, M.; Moses, M.A. Urinary TIMP-1 and MMP-2 levels detect the presence of pancreatic malignancies. *Br. J. Cancer* **2014**, *111*, 1772–1779. [CrossRef] [PubMed]

53. Xiang, T.; Xia, X.; Yan, W. Expression of Matrix Metalloproteinases-2/-9 is Associated with Microvessel Density in Pancreatic Cancer. *Am. J. Ther.* **2017**, *24*, e431–e434. [CrossRef] [PubMed]

54. Jakubowska, K.; Pryczynicz, A.; Januszewska, J.; Sidorkiewicz, I.; Kemona, A.; Niewiński, A.; Lewczuk, L.; Kędra, B.; Guzińska-Ustymowicz, K. Expressions of Matrix Metalloproteinases 2, 7, and 9 in Carcinogenesis of Pancreatic Ductal Adenocarcinoma. *Dis. Markers* **2016**, *2016*, 9895721. [CrossRef] [PubMed]

55. Singh, N.; Gupta, S.; Pandey, R.M.; Sahni, P.; Chauhan, S.S.; Saraya, A. Prognostic significance of plasma matrix metalloprotease-2 in pancreatic cancer patients. *Indian J. Med. Res.* **2017**, *146*, 334–340. [PubMed]

56. Mao, L.; Le, S.; Jin, X.; Liu, G.; Chen, J.; Hu, J. CSN5 promotes the invasion and metastasis of pancreatic cancer by stabilization of FOXM1. *Exp. Cell Res.* **2019**, *374*, 274–281. [CrossRef]

57. Li, Y.; Song, T.; Chen, Z.; Wang, Y.; Zhang, J.; Wang, X. Pancreatic Stellate Cells Activation and Matrix Metallopeptidase 2 Expression Correlate With Lymph Node Metastasis in Pancreatic Carcinoma. *Am. J. Med. Sci.* **2019**, *357*, 16–22. [CrossRef]

58. Zhu, C.L.; Huang, Q. Overexpression of the SMYD3 Promotes Proliferation, Migration, and Invasion of Pancreatic Cancer. *Dig. Dis. Sci.* **2020**, *65*, 489–499. [CrossRef]

59. Gress, T.M.; Müller-Pillasch, F.; Lerch, M.M.; Friess, H.; Büchler, M.; Adler, G. Expression and in-situ localization of genes coding for extracellular matrix proteins and extracellular matrix degrading proteases in pancreatic cancer. *Int. J. Cancer* **1995**, *62*, 407–413. [CrossRef]

60. Koshiba, T.; Hosotani, R.; Wada, M.; Fujimoto, K.; Lee, J.U.; Doi, R.; Arii, S.; Imamura, M. Detection of matrix metalloproteinase activity in human pancreatic cancer. *Surg. Today* **1997**, *27*, 302–304. [CrossRef]

61. Koshiba, T.; Hosotani, R.; Wada, M.; Miyamoto, Y.; Fujimoto, K.; Lee, J.U.; Doi, R.; Arii, S.; Imamura, M. Involvement of matrix metalloproteinase-2 activity in invasion and metastasis of pancreatic carcinoma. *Cancer* **1998**, *82*, 642–650. [CrossRef]

62. Kuniyasu, H.; Ellis, L.M.; Evans, D.B.; Abbruzzese, J.L.; Fenoglio, C.J.; Bucana, C.D.; Cleary, K.R.; Tahara, E.; Fidler, I.J. Relative expression of E-cadherin and type IV collagenase genes predicts disease outcome in patients with resectable pancreatic carcinoma. *Clin. Cancer Res.* **1999**, *5*, 25–33. [PubMed]

63. Pryczynicz, A.; Guzińska-Ustymowicz, K.; Dymicka-Piekarska, V.; Czyzewska, J.; Kemona, A. Expression of matrix metalloproteinase 9 in pancreatic ductal carcinoma is associated with tumor metastasis formation. *Folia Histochem. Cytobiol.* **2007**, *45*, 37–40. [PubMed]

64. Tian, M.; Cui, Y.Z.; Song, G.H.; Zong, M.J.; Zhou, X.Y.; Chen, Y.; Han, J.X. Proteomic analysis identifies MMP-9, DJ-1 and A1BG as overexpressed proteins in pancreatic juice from pancreatic ductal adenocarcinoma patients. *BMC Cancer* **2008**, *8*, 241. [CrossRef]

65. Mroczko, B.; Lukaszewicz-Zajac, M.; Wereszczynska-Siemiatkowska, U.; Groblewska, M.; Gryko, M.; Kedra, B.; Jurkowska, G.; Szmitkowski, M. Clinical significance of the measurements of serum matrix metalloproteinase-9 and its inhibitor (tissue inhibitor of metalloproteinase-1) in patients with pancreatic cancer: Metalloproteinase-9 as an independent prognostic factor. *Pancreas* **2009**, *38*, 613–618. [CrossRef]

66. Duan, L.; Hu, X.Q.; Feng, D.Y.; Lei, S.Y.; Hu, G.H. GPC-1 may serve as a predictor of perineural invasion and a prognosticator of survival in pancreatic cancer. *Asian J. Surg.* **2013**, *36*, 7–12. [CrossRef]

67. Xu, Y.; Li, Z.; Jiang, P.; Wu, G.; Chen, K.; Zhang, X.; Li, X. The co-expression of MMP-9 and Tenascin-C is significantly associated with the progression and prognosis of pancreatic cancer. *Diagn. Pathol.* **2015**, *10*, 211. [CrossRef]

68. Łukaszewicz-Zając, M.; Gryko, M.; Pączek, S.; Szmitkowski, M.; Kędra, B.; Mroczko, B. Matrix metalloproteinase 2 (MMP-2) and its tissue inhibitor 2 (TIMP-2) in pancreatic cancer (PC). *Oncotarget* **2019**, *10*, 395–403. [CrossRef]

69. Park, H.D.; Kang, E.S.; Kim, J.W.; Lee, K.T.; Lee, K.H.; Park, Y.S.; Park, J.O.; Lee, J.; Heo, J.S.; Choi, S.H. Serum CA19-9, cathepsin D, and matrix metalloproteinase-7 as a diagnostic panel for pancreatic ductal adenocarcinoma. *Proteomics* **2012**, *12*, 3590–3597. [CrossRef]

70. Määttä, M.; Soini, Y.; Liakka, A.; Autio-Harmainen, H. Differential expression of matrix metalloproteinase (MMP)-2, MMP-9, and membrane type 1-MMP in hepatocellular and pancreatic adenocarcinoma: Implications for tumor progression and clinical prognosis. *Clin. Cancer Res.* **2000**, *6*, 2726–2734.

71. Pezzilli, R.; Corsi, M.M.; Barassi, A.; Morselli-Labate, A.M.; Dogliotti, G.; Casadei, R.; Corinaldesi, R.; D'Eril, G.M. The role of inflammation in patients with intraductal mucinous neoplasm of the pancreas and in those with pancreatic adenocarcinoma. *Anticancer Res.* **2010**, *30*, 3801–3805. [PubMed]

72. Nagakawa, Y.; Aoki, T.; Kasuya, K.; Tsuchida, A.; Koyanagi, Y. Histologic features of venous invasion, expression of vascular endothelial growth factor and matrix metalloproteinase-2 and matrix metalloproteinase-9, and the relation with liver metastasis in pancreatic cancer. *Pancreas* **2002**, *24*, 169–178. [CrossRef] [PubMed]

73. Park, J.K.; Kim, M.A.; Ryu, J.K.; Yoon, Y.B.; Kim, S.W.; Han, H.S.; Kang, G.H.; Kim, H.; Hwang, J.H.; Kim, Y.T. Postoperative prognostic predictors of pancreatic ductal adenocarcinoma: Clinical analysis and immunoprofile on tissue microarrays. *Ann. Surg. Oncol.* **2012**, *19*, 2664–2672. [CrossRef] [PubMed]

74. Haq, M.; Shafii, A.; Zervos, E.E.; Rosemurgy, A.S. Addition of matrix metalloproteinase inhibition to conventional cytotoxic therapy reduces tumor implantation and prolongs survival in a murine model of human pancreatic cancer. *Cancer Res.* **2000**, *60*, 3207–3211. [PubMed]

75. Zervos, E.E.; Norman, J.G.; Gower, W.R.; Franz, M.G.; Rosemurgy, A.S. Matrix metalloproteinase inhibition attenuates human pancreatic cancer growth in vitro and decreases mortality and tumorigenesis in vivo. *J. Surg. Res.* **1997**, *69*, 367–371. [CrossRef]

76. Zervox, E.E.; Franz, M.G.; Salhab, K.F.; Shafii, A.E.; Menendez, J.; Gower, W.R.; Rosemurgy, A.S. Matrix metalloproteinase inhibition improves survival in an orthotopic model of human pancreatic cancer. *J. Gastrointest. Surg.* **2000**, *4*, 614–619. [CrossRef]

77. Zervos, E.E.; Shafii, A.E.; Rosemurgy, A.S. Matrix metalloproteinase (MMP) inhibition selectively decreases type II MMP activity in a murine model of pancreatic cancer. *J. Surg. Res.* **1999**, *81*, 65–68. [CrossRef]

78. Jimenez, R.E.; Hartwig, W.; Antoniu, B.A.; Compton, C.C.; Warshaw, A.L.; Fernández-Del Castillo, C. Effect of matrix metalloproteinase inhibition on pancreatic cancer invasion and metastasis: An additive strategy for cancer control. *Ann. Surg.* **2000**, *231*, 644–654. [CrossRef]

79. Gao, C.C.; Gong, B.G.; Wu, J.B.; Cheng, P.G.; Xu, H.Y.; Song, D.K.; Li, F. MMI-166, a selective matrix metalloproteinase inhibitor, promotes apoptosis in human pancreatic cancer. *Med. Oncol.* **2015**, *32*, 418. [CrossRef]

80. Matsushita, A.; Onda, M.; Uchida, E.; Maekawa, R.; Yoshioka, T. Antitumor effect of a new selective matrix metalloproteinase inhibitor, MMI-166, on experimental pancreatic cancer. *Int. J. Cancer* **2001**, *92*, 434–440. [CrossRef]

81. Kilian, M.; Gregor, J.I.; Heukamp, I.; Hanel, M.; Ahlgrimm, M.; Schimke, I.; Kristiansen, G.; Ommer, A.; Walz, M.L.; Jacobi, C.A.; et al. Matrix metalloproteinase inhibitor RO 28-2653 decreases liver metastasis by reduction of MMP-2 and MMP-9 concentration in BOP-induced ductal pancreatic cancer in Syrian Hamsters: Inhibition of matrix metalloproteinases in pancreatic cancer. *Prostaglandins Leukot. Essent. Fatty Acids* **2006**, *75*, 429–434. [CrossRef] [PubMed]

82. Iki, K.; Tsutsumi, M.; Kido, A.; Sakitani, H.; Takahama, M.; Yoshimoto, M.; Motoyama, M.; Tatsumi, K.; Tsunoda, T.; Konishi, Y. Expression of matrix metalloproteinase 2 (MMP-2), membrane-type 1 MMP and tissue inhibitor of metalloproteinase 2 and activation of proMMP-2 in pancreatic duct adenocarcinomas in hamsters treated with N-nitrosobis(2-oxopropyl)amine. *Carcinogenesis* **1999**, *20*, 1323–1329. [CrossRef] [PubMed]

83. Andersson, P.; Yang, Y.; Hosaka, K.; Zhang, Y.; Fischer, C.; Braun, H.; Liu, S.; Yu, G.; Liu, S.; Beyaert, R.; et al. Molecular mechanisms of IL-33-mediated stromal interactions in cancer metastasis. *JCI Insight* **2018**, *3*, e122375. [CrossRef] [PubMed]

84. Koikawa, K.; Ohuchida, K.; Ando, Y.; Kibe, S.; Nakayama, H.; Takesue, S.; Endo, S.; Abe, T.; Okumura, T.; Iwamoto, C.; et al. Basement membrane destruction by pancreatic stellate cells leads to local invasion in pancreatic ductal adenocarcinoma. *Cancer Lett.* **2018**, *425*, 65–77. [CrossRef] [PubMed]

85. Lu, G.; Zheng, M.; Zhu, Y.; Sha, M.; Wu, Y.; Han, X. Selection of peptide inhibitor to matrix metalloproteinase-2 using phage display and its effects on pancreatic cancer cell lines PANC-1 and CFPAC-1. *Int. J. Biol. Sci.* **2012**, *8*, 650–662. [CrossRef]

86. Arnold, S.; Mira, E.; Muneer, S.; Korpanty, G.; Beck, A.W.; Holloway, S.E.; Mañes, S.; Brekken, R.A. Forced expression of MMP9 rescues the loss of angiogenesis and abrogates metastasis of pancreatic tumors triggered by the absence of host SPARC. *Exp. Biol. Med. (Maywood)* **2008**, *233*, 860–873. [CrossRef]

87. Awasthi, N.; Mikels-Vigdal, A.J.; Stefanutti, E.; Schwarz, M.A.; Monahan, S.; Smith, V.; Schwarz, R.E. Therapeutic efficacy of anti-MMP9 antibody in combination with nab-paclitaxel-based chemotherapy in pre-clinical models of pancreatic cancer. *J. Cell Mol. Med.* **2019**, *23*, 3878–3887. [CrossRef]

88. Bausch, D.; Pausch, T.; Krauss, T.; Hopt, U.T.; Fernandez-del-Castillo, C.; Warshaw, A.L.; Thayer, S.P.; Keck, T. Neutrophil granulocyte derived MMP-9 is a VEGF independent functional component of the angiogenic switch in pancreatic ductal adenocarcinoma. *Angiogenesis* **2011**, *14*, 235–243. [CrossRef]

89. Nakamura, T.; Kuwai, T.; Kim, J.S.; Fan, D.; Kim, S.J.; Fidler, I.J. Stromal metalloproteinase-9 is essential to angiogenesis and progressive growth of orthotopic human pancreatic cancer in parabiont nude mice. *Neoplasia* **2007**, *9*, 979–986. [CrossRef]

90. Grünwald, B.; Vandooren, J.; Gerg, M.; Ahomaa, K.; Hunger, A.; Berchtold, S.; Akbareian, S.; Schaten, S.; Knolle, P.; Edwards, D.R.; et al. Systemic Ablation of MMP-9 Triggers Invasive Growth and Metastasis of Pancreatic Cancer via Deregulation of IL6 Expression in the Bone Marrow. *Mol. Cancer Res.* **2016**, *14*, 1147–1158. [CrossRef]

91. Bournet, B.; Pointreau, A.; Souque, A.; Muscari, F.; Lepage, B.; Senesse, P.; Barthet, M.; Lesavre, N.; Hammel, P.; Levy, P.; et al. Gene expression signature of advanced pancreatic ductal adenocarcinoma using low density array on endoscopic ultrasound-guided fine needle aspiration samples. *Pancreatology* **2012**, *12*, 27–34. [CrossRef] [PubMed]

92. Lee, J.; Lee, J.; Kim, J.H. Identification of Matrix Metalloproteinase 11 as a Prognostic Biomarker in Pancreatic Cancer. *Anticancer Res.* **2019**, *39*, 5963–5971. [CrossRef] [PubMed]

93. Von Marschall, Z.; Riecken, E.O.; Rosewicz, S. Stromelysin 3 is overexpressed in human pancreatic carcinoma and regulated by retinoic acid in pancreatic carcinoma cell lines. *Gut* **1998**, *43*, 692–698. [CrossRef] [PubMed]

94. Mehner, C.; Miller, E.; Khauv, D.; Nassar, A.; Oberg, A.L.; Bamlet, W.R.; Zhang, L.; Waldmann, J.; Radisky, E.S.; Crawford, H.C.; et al. Tumor cell-derived MMP3 orchestrates Rac1b and tissue alterations that promote pancreatic adenocarcinoma. *Mol. Cancer Res.* **2014**, *12*, 1430–1439. [CrossRef] [PubMed]

95. Mehner, C.; Miller, E.; Nassar, A.; Bamlet, W.R.; Radisky, E.S.; Radisky, D.C. Tumor cell expression of MMP3 as a prognostic factor for poor survival in pancreatic, pulmonary, and mammary carcinoma. *Genes Cancer* **2015**, *6*, 480–489.

96. Zhang, J.J.; Zhu, Y.; Xie, K.L.; Peng, Y.P.; Tao, J.Q.; Tang, J.; Li, Z.; Xu, Z.K.; Dai, C.C.; Qian, Z.Y.; et al. Yin Yang-1 suppresses invasion and metastasis of pancreatic ductal adenocarcinoma by downregulating MMP10 in a MUC4/ErbB2/p38/MEF2C-dependent mechanism. *Mol. Cancer.* **2014**, *13*, 130. [CrossRef]

97. Fukushima, H.; Yamamoto, H.; Itoh, F.; Nakamura, H.; Min, Y.; Horiuchi, S.; Iku, S.; Sasaki, S.; Imai, K. Association of matrilysin mRNA expression with K-ras mutations and progression in pancreatic ductal adenocarcinomas. *Carcinogenesis* **2001**, *22*, 1049–1052. [CrossRef]

98. Crawford, H.C.; Scoggins, C.R.; Washington, M.K.; Matrisian, L.M.; Leach, S.D. Matrix metalloproteinase-7 is expressed by pancreatic cancer precursors and regulates acinar-to-ductal metaplasia in exocrine pancreas. *J. Clin. Investig.* **2002**, *109*, 1437–1444. [CrossRef]

99. Li, Y.J.; Wei, Z.M.; Meng, Y.X.; Ji, X.R. Beta-catenin up-regulates the expression of cyclinD1, c-myc and MMP-7 in human pancreatic cancer: Relationships with carcinogenesis and metastasis. *World J. Gastroenterol.* **2005**, *11*, 2117–2123. [CrossRef]

100. Kuhlmann, K.F.; van Till, J.W.; Boermeester, M.A.; de Reuver, P.R.; Tzvetanova, I.D.; Offerhaus, G.J.; ten Kate, F.J.; Busch, O.R.; van Gulik, T.M.; Gouma, D.J.; et al. Evaluation of matrix metalloproteinase 7 in plasma and pancreatic juice as a biomarker for pancreatic cancer. *Cancer Epidemiol. Biomark. Prev.* **2007**, *16*, 886–891. [CrossRef]

101. Yuan, P.; He, X.H.; Rong, Y.F.; Cao, J.; Li, Y.; Hu, Y.P.; Liu, Y.; Li, D.; Lou, W.; Liu, M.F. KRAS/NF-κB/YY1/miR-489 Signaling Axis Controls Pancreatic Cancer Metastasis. *Cancer Res.* **2017**, *77*, 100–111. [CrossRef] [PubMed]

102. Banaei, N.; Foley, A.; Houghton, J.M.; Sun, Y.; Kim, B. Multiplex detection of pancreatic cancer biomarkers using a SERS-based immunoassay. *Nanotechnology* **2017**, *28*, 455101. [CrossRef] [PubMed]

103. Resovi, A.; Bani, M.R.; Porcu, L.; Anastasia, A.; Minoli, L.; Allavena, P.; Cappello, P.; Novelli, F.; Scarpa, A.; Morandi, E.; et al. Soluble stroma-related biomarkers of pancreatic cancer. *EMBO Mol. Med.* **2018**, *10*, e8741. [CrossRef] [PubMed]

104. Yang, J.; Cong, X.; Ren, M.; Sun, H.; Liu, T.; Chen, G.; Wang, Q.; Li, Z.; Yu, S.; Yang, Q. Circular RNA hsa_circRNA_0007334 is Predicted to Promote MMP7 and COL1A1 Expression by Functioning as a miRNA Sponge in Pancreatic Ductal Adenocarcinoma. *J. Oncol.* **2019**, *2019*, 7630894. [CrossRef] [PubMed]

105. Fukuda, A.; Wang, S.C.; Morris, J.P., IV; Folias, A.E.; Liou, A.; Kim, G.E.; Akira, S.; Boucher, K.M.; Firpo, M.A.; Mulvihill, S.J.; et al. Stat3 and MMP7 contribute to pancreatic ductal adenocarcinoma initiation and progression. *Cancer Cell* **2011**, *19*, 441–455. [CrossRef] [PubMed]

106. Bister, V.; Skoog, T.; Virolainen, S.; Kiviluoto, T.; Puolakkainen, P.; Saarialho-Kere, U. Increased expression of matrix metalloproteinases-21 and -26 and TIMP-4 in pancreatic adenocarcinoma. *Mod. Pathol.* **2007**, *20*, 1128–1140. [CrossRef]

107. Krantz, S.B.; Shields, M.A.; Dangi-Garimella, S.; Cheon, E.C.; Barron, M.R.; Hwang, R.F.; Rao, M.R.; Grippo, P.J.; Bentrem, D.J.; Munshi, H.G. MT1-MMP cooperates with Kras(G12D) to promote pancreatic fibrosis through increased TGF-β signaling. *Mol. Cancer Res.* **2011**, *9*, 1294–1304. [CrossRef]

108. Dangi-Garimella, S.; Krantz, S.B.; Barron, M.R.; Shields, M.A.; Heiferman, M.J.; Grippo, P.J.; Bentrem, D.J.; Munshi, H.G. Three-dimensional collagen I promotes gemcitabine resistance in pancreatic cancer through MT1-MMP-mediated expression of HMGA2. *Cancer Res.* **2011**, *71*, 1019–1028. [CrossRef]

109. Qiang, L.; Cao, H.; Chen, J.; Weller, S.G.; Krueger, E.W.; Zhang, L.; Razidlo, G.L.; McNiven, M.A. Pancreatic tumor cell metastasis is restricted by MT1-MMP binding protein MTCBP-1. *J. Cell Biol.* **2019**, *218*, 317–332. [CrossRef]

110. Ottaviano, A.J.; Sun, L.; Ananthanarayanan, V.; Munshi, H.G. Extracellular matrix-mediated membrane-type 1 matrix metalloproteinase expression in pancreatic ductal cells is regulated by transforming growth factor-beta1. *Cancer Res.* **2006**, *66*, 7032–7040. [CrossRef]

111. Balaz, P.; Friess, H.; Kondo, Y.; Zhu, Z.; Zimmermann, A.; Büchler, M.W. Human macrophage metalloelastase worsens the prognosis of pancreatic cancer. *Ann. Surg.* **2002**, *235*, 519–527. [CrossRef] [PubMed]

112. Zhai, L.L.; Wu, Y.; Cai, C.Y.; Huang, Q.; Tang, Z.G. High-Level Expression and Prognostic Significance of Matrix Metalloprotease-19 and Matrix Metalloprotease-20 in Human Pancreatic Ductal Adenocarcinoma. *Pancreas* **2016**, *45*, 1067–1072. [CrossRef] [PubMed]

113. Khalid, M.; Idichi, T.; Seki, N.; Wada, M.; Yamada, Y.; Fukuhisa, H.; Toda, H.; Kita, Y.; Kawasaki, Y.; Tanoue, K.; et al. Gene Regulation by Antitumor miR-204-5p in Pancreatic Ductal Adenocarcinoma: The Clinical Significance of Direct RACGAP1 Regulation. *Cancers* **2019**, *11*, 327. [CrossRef] [PubMed]

114. Bramhall, S.R.; Schulz, J.; Nemunaitis, J.; Brown, P.D.; Baillet, M.; Buckels, J.A. A double-blind placebo-controlled, randomised study comparing gemcitabine and marimastat with gemcitabine and placebo as first line therapy in patients with advanced pancreatic cancer. *Br. J. Cancer* **2002**, *87*, 161–167. [CrossRef]

115. Moore, M.J.; Hamm, J.; Dancey, J.; Dagenais, M.; Fields, A.; Hagan, K.; Greenberg, B.; Colwell, B.; Zee, B.; Tuet, D.; et al. Comparison of gemcitabine versus the matrix metalloproteinase inhibitor BAY 12-9566 in patients with advanced or metastatic adenocarcinoma of the pancreas: A phase III trial of the National Cancer Institute of Canada Clinical Trials Group. *J. Clin. Oncol.* **2003**, *21*, 3296–3302. [CrossRef]

116. Fields, G.B. The Rebirth of Matrix Metalloproteinase Inhibitors: Moving Beyond the Dogma. *Cells* **2019**, *8*, 984. [CrossRef]

117. Winer, A.; Adams, S.; Mignatti, P. Matrix Metalloproteinase Inhibitors in Cancer Therapy: Turning Past Failures into Future Successes. *Mol. Cancer Ther.* **2018**, *17*, 1147–1155. [CrossRef]

118. Fujimoto, N.; Mouri, N.; Iwata, K.; Ohuchi, E.; Okada, Y.; Hayakawa, T. A one-step sandwich enzyme immunoassay for human matrix metalloproteinase 2 (72-kDa gelatinase/type IV collagenase) using monoclonal antibodies. *Clin. Chim. Acta* **1993**, *221*, 91–103. [CrossRef]

119. Lin, F.; Wang, X.; Jie, Z.; Hong, X.; Li, X.; Wang, M.; Yu, Y. Inhibitory effects of miR-146b-5p on cell migration and invasion of pancreatic cancer by targeting MMP16. *J. Huazhong Univ. Sci. Technolog. Med. Sci.* **2011**, *31*, 509. [CrossRef]

120. Giannopoulos, G.; Pavlakis, K.; Parasi, A.; Kavatzas, N.; Tiniakos, D.; Karakosta, A.; Tzanakis, N.; Peros, G. The expression of matrix metalloproteinases-2 and -9 and their tissue inhibitor 2 in pancreatic ductal and ampullary carcinoma and their relation to angiogenesis and clinicopathological parameters. *Anticancer Res.* **2008**, *28*, 1875–1881.

121. Lee, S.H.; Kim, H.; Hwang, J.H.; Shin, E.; Lee, H.S.; Hwang, D.W.; Cho, J.Y.; Yoon, Y.S.; Han, H.S.; Cha, B.H. CD24 and S100A4 expression in resectable pancreatic cancers with earlier disease recurrence and poor survival. *Pancreas* **2014**, *43*, 380–388. [CrossRef] [PubMed]

122. Wang, S.C.; Parekh, J.R.; Porembka, M.R.; Nathan, H.; D'Angelica, M.I.; DeMatteo, R.P.; Fong, Y.; Kingham, T.P.; Jarnagin, W.R.; Allen, P.J. A Pilot Study Evaluating Serum MMP7 as a Preoperative Prognostic Marker for Pancreatic Ductal Adenocarcinoma Patients. *J. Gastrointest. Surg.* **2016**, *20*, 899–904. [CrossRef] [PubMed]

123. Nakamura, H.; Horita, S.; Senmaru, N.; Miyasaka, Y.; Gohda, T.; Inoue, Y.; Fujita, M.; Meguro, T.; Morita, T.; Nagashima, K. Association of matrilysin expression with progression and poor prognosis in human pancreatic adenocarcinoma. *Oncol. Rep.* **2002**, *9*, 751–755. [CrossRef] [PubMed]

124. Harvey, S.R.; Hurd, T.C.; Markus, G.; Martinick, M.I.; Penetrante, R.M.; Tan, D.; Venkataraman, P.; DeSouza, N.; Sait, S.N.; Driscoll, D.L.; et al. Evaluation of urinary plasminogen activator, its receptor, matrix metalloproteinase-9, and von Willebrand factor in pancreatic cancer. *Clin. Cancer Res.* **2003**, *9*, 4935–4943. [PubMed]

125. Cho, K.; Matsuda, Y.; Ueda, J.; Uchida, E.; Naito, Z.; Ishiwata, T. Keratinocyte growth factor induces matrix metalloproteinase-9 expression and correlates with venous invasion in pancreatic cancer. *Int. J. Oncol.* **2012**, *40*, 1040–1048. [CrossRef] [PubMed]
126. Liu, Q.; Sheng, W.; Dong, M.; Dong, X.; Dong, Q.; Li, F. Gli1 promotes transforming growth factor-beta1- and epidermal growth factor-induced epithelial to mesenchymal transition in pancreatic cancer cells. *Surgery* **2015**, *158*, 211–224. [CrossRef] [PubMed]
127. Zhu, S.K.; Zhou, Y.; Cheng, C.; Zhong, S.; Wu, H.Q.; Wang, B.; Fan, P.; Xiong, J.X.; Yang, H.J.; Wu, H.S. Overexpression of membrane-type 2 matrix metalloproteinase induced by hypoxia-inducible factor-1α in pancreatic cancer: Implications for tumor progression and prognosis. *Mol. Clin. Oncol.* **2014**, *2*, 973–981. [CrossRef]
128. Gopinathan, A.; Morton, J.P.; Jodrell, D.; Sansom, O.J. GEMMs as preclinical models for testing pancreatic cancer therapies. *Dis. Model. Mech.* **2015**, *8*, 1185–1200. [CrossRef]
129. Swayden, M.; Soubeyran, P.; Iovanna, J. Upcoming Revolutionary Paths in Preclinical Modeling of Pancreatic Adenocarcinoma. *Front. Oncol.* **2020**, *9*, 1443. [CrossRef]
130. Biankin, A.V.; Waddell, N.; Kassahn, K.S.; Gingras, M.C.; Muthuswamy, L.B.; Johns, A.L.; Miller, D.K.; Wilson, P.J.; Patch, A.M.; Wu, J.; et al. Pancreatic cancer genomes reveal aberrations in axon guidance pathway genes. *Nature* **2012**, *491*, 399–405. [CrossRef]

Review

Proteolytic Cleavages in the VEGF Family: Generating Diversity among Angiogenic VEGFs, Essential for the Activation of Lymphangiogenic VEGFs

Jaana Künnapuu [1], Honey Bokharaie [1] and Michael Jeltsch [1,2,3,*]

1 Drug Research Program, Faculty of Pharmacy, University of Helsinki, 00014 Helsinki, Finland; jaana.vulli@helsinki.fi (J.K.); honey.bokharaie@helsinki.fi (H.B.)
2 Individualized Drug Therapy Research Program, Faculty of Medicine, University of Helsinki, 00014 Helsinki, Finland
3 Wihuri Research Institute, 00290 Helsinki, Finland
* Correspondence: michael@jeltsch.org; Tel.: +358-50-3200235

Simple Summary: Vascular endothelial growth factors (VEGFs) regulate the growth of blood and lymphatic vessels. Some of them induce the growth of blood vessels, and others the growth of lymphatic vessels. Blocking VEGF-A is used today to treat several types of cancer ("antiangiogenic therapy"). However, in other diseases, we would like to increase the activity of VEGFs. For example, VEGF-A could generate new blood vessels to protect from heart disease, and VEGF-C could generate new lymphatics to counteract lymphedema. Clinical trials are testing the latter concept at the moment. Because VEGF-C and VEGF-D are produced as inactive precursors, we propose that novel drugs could also target the enzymatic activation of VEGF-C and VEGF-D. However, because of the delicate balance between too much and too little vascular growth, a detailed understanding of the activation of the VEGFs is needed before such concepts can be converted into safe and efficacious therapies.

Abstract: Specific proteolytic cleavages turn on, modify, or turn off the activity of vascular endothelial growth factors (VEGFs). Proteolysis is most prominent among the lymphangiogenic VEGF-C and VEGF-D, which are synthesized as precursors that need to undergo enzymatic removal of their C- and N-terminal propeptides before they can activate their receptors. At least five different proteases mediate the activating cleavage of VEGF-C: plasmin, ADAMTS3, prostate-specific antigen, cathepsin D, and thrombin. All of these proteases except for ADAMTS3 can also activate VEGF-D. Processing by different proteases results in distinct forms of the "mature" growth factors, which differ in affinity and receptor activation potential. The "default" VEGF-C-activating enzyme ADAMTS3 does not activate VEGF-D, and therefore, VEGF-C and VEGF-D do function in different contexts. VEGF-C itself is also regulated in different contexts by distinct proteases. During embryonic development, ADAMTS3 activates VEGF-C. The other activating proteases are likely important for non-developmental lymphangiogenesis during, e.g., tissue regeneration, inflammation, immune response, and pathological tumor-associated lymphangiogenesis. The better we understand these events at the molecular level, the greater our chances of developing successful therapies targeting VEGF-C and VEGF-D for diseases involving the lymphatics such as lymphedema or cancer.

Keywords: vascular endothelial growth factors (VEGFs); VEGF-A; PlGF; VEGF-B; VEGF-C; VEGF-D; angiogenesis; lymphangiogenesis; CCBE1; proteases; ADAMTS3; plasmin; cathepsin D; KLK3; prostate-specific antigen (PSA); thrombin; wound healing; metastasis; proteolytic activation; vascular biology; lymphedema

check for updates

Citation: Künnapuu, J.; Bokharaie, H.; Jeltsch, M. Proteolytic Cleavages in the VEGF Family: Generating Diversity among Angiogenic VEGFs, Essential for the Activation of Lymphangiogenic VEGFs. *Biology* **2021**, *10*, 167. https://doi.org/10.3390/biology10020167

Academic Editor: Hang Fai Kwok

Received: 15 January 2021
Accepted: 18 February 2021
Published: 23 February 2021

Publisher's Note: MDPI stays neutral with regard to jurisdictional claims in published maps and institutional affiliations.

1. Introduction

In vertebrates, the family of vascular endothelial growth factors (VEGFs) typically comprises five genes: VEGF-A (in older literature often referred to simply as "VEGF"), placenta growth factor (PlGF), VEGF-B, VEGF-C, and VEGF-D. In addition to these orthodox VEGFs, several genes coding for VEGF-like molecules have been discovered in some members of the poxvirus and iridovirus families (collectively named VEGF-E) [1–4] and in venomous reptiles (collectively named VEGF-F) [5]. In vertebrates, the VEGF growth factors are central to the development and maintenance of the cardiovascular system and the lymphatic system. Non-vertebrates also feature VEGF-like molecules, but their functions are less well defined.

The subdivision of the vertebrate vascular system into the cardiovascular and the lymphatic system is reflected at the molecular level by a subdivision of the VEGF family into VEGFs acting primarily on blood vessels (VEGF-A, PlGF, and VEGF-B) and VEGFs acting mostly on lymphatic vessels (VEGF-C and VEGF-D). This specificity results from the expression pattern of the three VEGF receptors (VEGFRs). VEGFR-1 and VEGFR-2 are expressed on blood vascular endothelial cells (BECs), while lymphatic endothelial cells (LECs) express VEGFR-2 and VEGFR-3 (Figure 1).

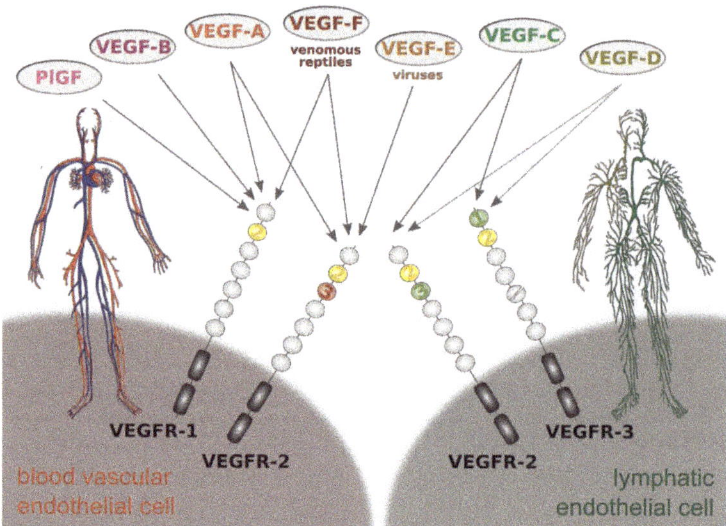

Figure 1. Vascular endothelial growth factors (VEGFs) act on blood vessels and/or lymphatic vessels depending on their affinities towards VEGF receptors 1, -2, and -3. VEGFR-2 is expressed on both blood and lymphatic endothelium. In principle, growth factors that do activate VEGFR-2 can promote both the growth of blood vessels (angiogenesis) and lymphatic vessels (lymphangiogenesis). VEGF-E and VEGF-F are not of human origin: VEGF-E genes are found in viral genomes, and VEGF-F is a snake venom component. All receptor-growth factor interactions require the extracellular domain 2 of the VEGF receptors (shown in yellow) [6–9]. Domain 3 of VEGFR-2 is important for the interaction of VEGFR-2 with both VEGF-A [7] and VEGF-C [8], and domain 1 of VEGFR-3 is important for the interaction of VEGF-C with VEGFR-3 [9].

The biology of the VEGFs and their signaling pathways has been extensively discussed elsewhere [10,11]. From all VEGF family members, only VEGF-A and VEGF-C are essential in the sense that constitutive ablation of their genes in mice results in embryonic lethality [12–14]. VEGF-A levels are so crucial that even heterozygous mice are not viable. In fact, *VEGFA* was the first gene where the deletion of a single allele was shown to be embryonically lethal [12,13]. While the primary function and importance of the cardiovascular system are also obvious to the layperson, the tasks of the lymphatic system escape even some life science professionals. Its major three tasks are:

1. Tissue drainage for fluid balance and waste disposal

2. Immune surveillance, including hosting and trafficking of immune cells
3. Uptake of dietary long-chain fatty acids and other highly lipophilic compounds in the intestine

Considerable effort has been devoted to the mechanisms and effects of receptor binding and downstream signaling of the VEGFs. Less is known about the processes upstream of receptor binding such as secretion, release, and proteolytic processing. In this review, we want to briefly give an overview of what is known about the proteolytic processing of VEGFs with a focus on the lymphangiogenic VEGFs.

Evolutionarily, the importance of proteases has been remarkable. Proteolytic processing often regulates protein activity and creates variation in a protein's function. This has been suggested by phylogenetic and functional studies in all kingdoms of life, including viruses [15], plants [16], and animals [17]. Not surprisingly, proteases are used to regulate function and create functional variety in the VEGF family and can be regarded as signaling molecules [18].

2. Proteolytic Processing of the Hemangiogenic VEGFs

Among the hemangiogenic VEGFs, protein diversification within a single VEGF family member relies more on differential mRNA splicing than on proteolytic processing (Figure 2; reviewed in [19]). mRNA splicing generates several isoforms of VEGF-A, which differ by the extent of the C-terminal, predominantly basic heparin-binding domain (HBD) [20–22]. The HBD mediates the interaction of VEGF-A with the extracellular matrix (ECM), cell surface heparan sulfate proteoglycans (HSPGs), and neuropilin-1. The interaction with HSPGs involves both a sequence-specific binding epitope and electrostatic effects of a predominantly basic amino acid sequence. Only a few isoforms are entirely devoid of heparin-binding properties under physiological conditions and therefore fully soluble. Mice expressing only the major soluble isoform (VEGF-A$_{121}$) are born but show severe cardiovascular defects and die from cardiac failure [23].

The matrix-binding properties of the larger VEGF-A isoforms are essential for generating growth factor gradients, which are assumed to be essential for efficient organ vascularization [24,25]. VEGF-A$_{189}$ and VEGF-A$_{206}$ are sequestered in the extracellular matrix (or on cell surface HSPGs), and at least VEGF-A$_{189}$ has been shown not to participate in receptor activation [26]. Proteases such as plasmin, urokinase-type plasminogen activator (uPA), and factor VII-activating protease (FSAP) can release and thus activate the ECM-bound, longer VEGF-A isoforms [21,27–29]. The cleavage of the primary isoform VEGF-A$_{165}$ can also be mediated by various matrix metalloproteinases (MMPs), especially MMP-3, resulting in smaller, non-heparin-binding products [30]. While such cleavages do liberate VEGF-A and are necessary for the mitogenic activity of VEGF-A$_{189}$ [26], they were reported to reduce the mitogenicity of VEGF-A$_{165}$ [31]. Unfortunately, there is little insight into the nature of the molecular handover of HSPG- and ECM-bound VEGF-A to VEGFR-2 or the VEGFR-2/neuropilin signaling complex [11]. The isoform composition and the location where the cleavage happens are likely important determinants of the net effect. The release of cell surface HSPG-bound VEGF-A is perhaps more likely to result in productive signaling than the release of ECM-bound VEGF-A. Complementary to the ECM release by proteolytic cleavage of VEGF-A, enzymatic degradation of the ECM-binding sites, e.g., of HSPGs by heparinases, or binding site competition by heparin or heparan sulfate achieves the same release, but without loss of the HBD [27].

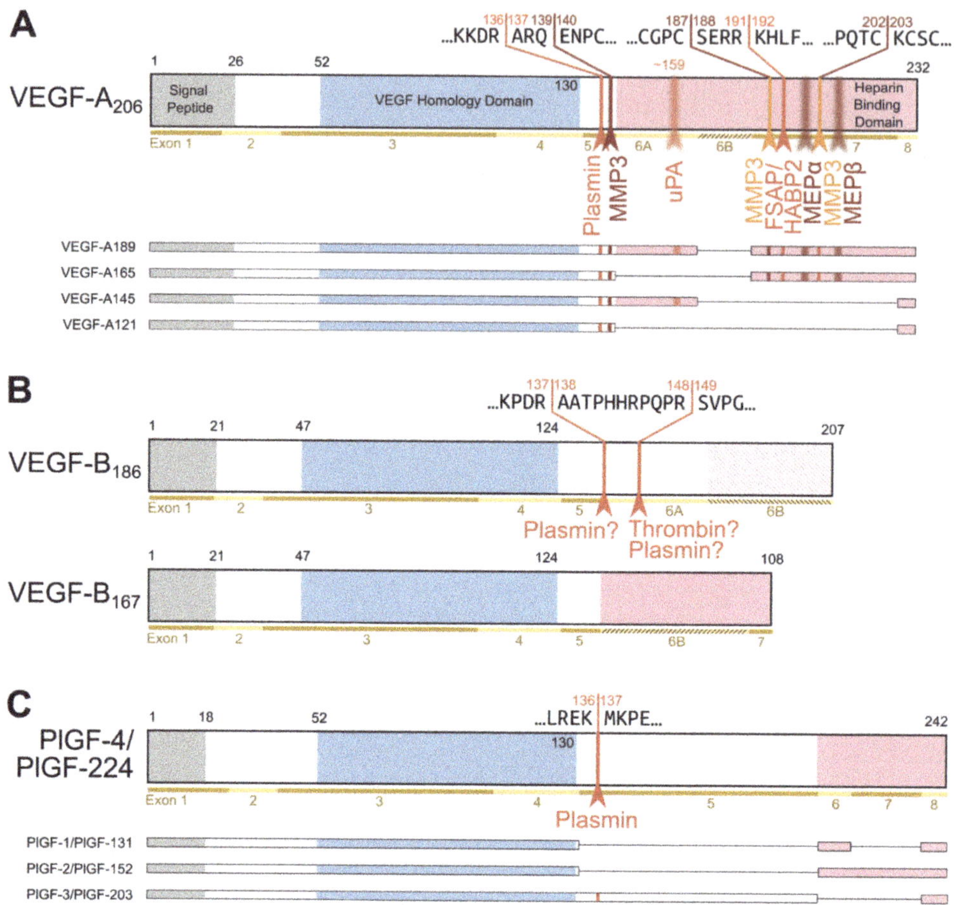

Figure 2. Most diversity among the hemangiogenic VEGFs is achieved by alternative splicing. Nevertheless, proteolytic processing of VEGF-A (**A**) [19] and placenta growth factor (PlGF) [32] (**C**) can convert the longer, heparin-binding isoforms into more soluble shorter species. (**B**) VEGF-B is a special case. Alternative splicing results in two isoforms that translate the same nucleotide sequence in two different frames resulting in a heparin-binding and a soluble isoform [33,34]. Due to the near-perfect cleavage context [35], thrombin has been suspected to be the responsible protease for VEGF-B$_{186}$ cleavage [36]. Prothrombin is indeed expressed by 293T cells [37], in which the cleavage has been demonstrated [33]. Plasmin cleaves VEGF-B$_{186}$ at at least four different sites, of which the two most likely predicted sites are indicated. Importantly, the predicted plasmin cleavage between Arg137 and Ala138 removes the interaction epitope for neuropilin-1 binding [33] Semi-transparent, blurry arrows indicate cleavages, for which only the approximate position is known. The figure shows only the most sensitive site from the plasmin cleavages of VEGF-A since prolonged incubation results in progressing degradation [30]. VEGF-B$_{186}$ appears to be progressively degraded by plasmin as well [33]. For VEGF-A and PlGF, the numbering is according to the longest shown isoform. VEGF-A is cleaved not only by MMP3 but also in a similar fashion by MMP7, MMP9, MMP19, and - less efficiently - by MMP1 and MMP16 [30].

Of the four human PlGF isoforms, PlGF-2 and -4 also contain a C-terminal heparin-binding domain. At least the PlGF-2 HBD can be removed by plasmin [32]. VEGF-B$_{167}$ also contains a heparin-binding domain homologous to the one in VEGF-A$_{165}$, but it is unknown whether this domain is subject to proteolytic removal. A yet unknown protease unmasks the neuropilin-1 binding site of the longer VEGF-B$_{186}$ isoform, but its target site is absent in VEGF-B$_{167}$ [33]. The cleavage context suggests that thrombin can unmask the neuropilin-1 binding epitope (see Figure 2) [35]. Plasmin cleavage at the same site is likely but does not result in neuropilin-1 binding due to additional cleavages that remove important sequences for neuropilin-1 binding [33].

3. The Lymphangiogenic Growth Factors VEGF-C and VEGF-D

The hemangiogenic VEGFs are rendered inactive either through ECM-association or—as in the case for VEGF-A$_{189}$—by their C-terminal auxiliary domain. Preventing receptor activation using inhibitory domains is also characteristic of the lymphangiogenic VEGFs. Upon secretion, VEGF-C and VEGF-D are kept inactive by their N- and C-terminal propeptides. Hence, the secreted forms are referred to as pro-VEGF-C and pro-VEGF-D. The removal of the propeptides requires two concerted proteolytic cleavages and happens in a very similar fashion for both VEGF-C and VEGF-D (see Figure 3):

1. Protein convertases constitutively cleave VEGF-C before secretion. This intracellular cleavage occurs between the central VEGF homology domain (VHD) and the C-terminal propeptide. However, it does not remove the C-terminal propeptide because it remains covalently attached to the rest of the molecule by disulfide bonds [38–40].

2. The second, extracellular cleavage activates the protein. This cleavage occurs between the N-terminal propeptide and the VHD [38] and can be mediated by different proteases. ADAMTS3 mediates VEGF-C activation in the embryonic development of the mammalian lymphatic system [41–43]. ADAMTS3 is specific for VEGF-C and does not activate VEGF-D. All other activating proteases target both VEGF-C and VEGF-D: plasmin [43,44], prostate-specific antigen (KLK3/PSA), cathepsin D (CatD) [45], and thrombin [46]. The resulting forms of VEGF-C and VEGF-D are referred to as active, mature, or short forms. However, they differ from each other at their N-termini because different proteases cleave at different positions within the linker between the N-terminal propeptide and the VHD (see Figure 4).

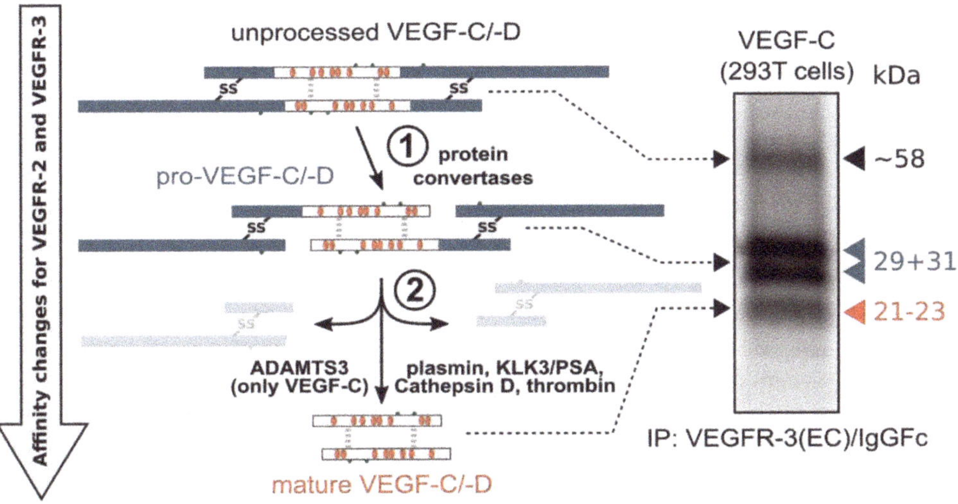

Figure 3. Two proteolytic cleavages are needed to activate VEGF-C and VEGF-D. The first cleavage, by protein convertases, is constitutive and intracellular. The second is highly regulated and happens after secretion of the pro-forms. Many different enzymes have been shown to catalyze the second cleavage, but the primary activating protease of VEGF-C in mammalian developmental lymphangiogenesis is A Disintegrin and Metalloprotease With Thrombospondin Motifs-3 (ADAMTS3). The immunoprecipitation (IP) of transfected 293T cells with a VEGFR-3(EC)/IgGFc fusion protein pulls down the 58 kDa full-length VEGF-C, the pro-VEGF-C peptides of 31 kDa and 29 kDa, and the mature VEGF-C. Proteins were resolved under reducing conditions by SDS-PAGE.

Interestingly, pro-VEGF-C can competitively block the receptor activation of active, mature VEGF-C. Its propeptides allow VEGF receptor binding but interfere with receptor activation. Apart from VEGFR-3, pro-VEGF-C also binds the co-receptor neuropilin-2. C-terminal propeptide processing exposes two terminal arginines (R226,227), which contribute to the conserved binding site for neuropilins [47]. Because it is not entirely clear whether pro-VEGF-C is completely incapable of receptor activation or whether it has some residual activity, pro-VEGF-C is either a partial agonist or an antagonist of mature VEGF-C [43].

4. Plasmin and Thrombin

The serine protease plasmin was the first protease that was shown to activate both VEGF-C and VEGF-D. Plasmin can remove both the N- and the C-terminal propeptides of VEGF-D to create a mature form containing only the VEGF homology domain [44]. One of plasmin's main functions is to degrade fibrin, the main component of blood clots. Thrombin is the newest addition to the group of VEGF-C/D-activating enzymes. In addition to its classical role in converting soluble fibrinogen into insoluble fibrin fibrils during blood clotting, it plays a crucial role in early wound healing [46] by activating VEGF-C, which is released from α-granules upon platelet aggregation [48]. Hence both thrombin and plasmin act concertedly to maintain a supply of active VEGF-C over the entire wound healing period.

However, in vitro, where no feedback loop exists to limit plasmin activity, prolonged exposure of VEGF-C results in VEGF-C inactivation [43]. In any case, without tissue damage, inactive prothrombin is not converted into thrombin, and inactive plasminogen not into plasmin. Therefore, in vivo, VEGF-C activation by thrombin or plasmin is likely restricted to situations with tissue damage. Using a similar rationale, platelet-rich plasma has been proposed for the treatment of lymphedema [49] and to promote wound healing [50]. Some proteomics studies have occasionally missed VEGF-C (as well as VEGF-A) when examining the platelet proteome, which might result from the relative resistance of the VEGF cystine knot to digestion with trypsin or similar proteases (unpublished data by the author). Nevertheless, other analyses and pharmacokinetic studies on anti-VEGF-C antibodies confirm the early findings of VEGF-C release during blood coagulation [51,52].

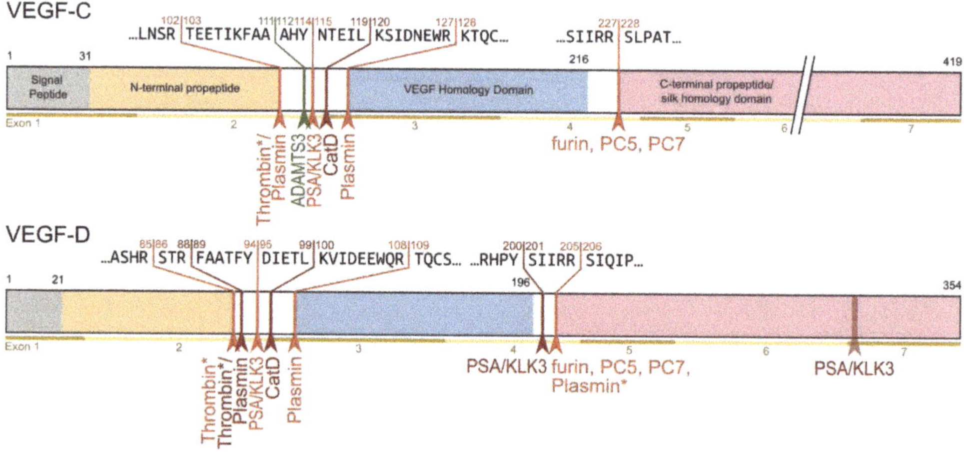

Figure 4. Human VEGF-C and -D are processed in a very similar fashion. The major difference between VEGF-C and -D is that ADAMTS3 activates VEGF-C, but not VEGF-D. This is one of the reasons why ADAMTS3 and VEGF-C are essential for lymphatic development and embryonic survival [14,42], whereas VEGF-D deletion in mice is well tolerated [53]. While the figure shows the exon structure of VEGF-C and -D, mRNA splice isoforms have only been reported for murine Vegfc [54]. The detected splice variants do not contain the full VEGF homology domain and are therefore not shown here. *Cleavage site is only predicted based on the amino acid context.

Plasmin activation of VEGF-C, which has been shown independently by two different groups [43,44], was not detected in a recent study [46]. Possibly, cleavage products might not have been recognized by the antibody due to low sensitivity or an absent epitope. Alternatively, the internal FLAG-tag preceding the cleavage site, which was used to prevent detection failure due to isoform-specific VEGF-C antibodies, might have interfered with the activation.

5. ADAMTS3 and the Cofactor CCBE1

ADAMTS3 was identified in the search for the endogenous protease that activates VEGF-C. Although plasmin had been identified as a VEGF-C-activating protease [44], it was never seriously considered as a physiological activator of VEGF-C due to its function in fibrin clot degradation. Moreover, lymphatic phenotypes have never been reported for the plasminogen knock-out mice [55] or human homozygous functional ablations [56].

In 2009, Alders used homozygosity mapping to identify mutations in the human *CCBE1* gene as a cause of Hennekam Syndrome (HS) [57], which is characterized by generalized lymphatic dysplasia [58]. When the same *Ccbe1* gene was ablated in zebrafish or mice [59,60], the phenotype was closely phenocopying the *Vegfc* knock-out [14]. Because it lacks any protease signature, CCBE1 was assumed to be somehow essential for the VEGF-C/VEGFR-3 signaling pathway, but not to be the VEGF-C-activating protease itself.

Co-transfection of CCBE1 with VEGF-C demonstrated that CCBE1 enhances the proteolytic processing of VEGF-C in 293T cells, and ADAMTS3 was identified as the responsible protease by mass spectrometric analysis of a partially purified CCBE1 from a CCBE1-overexpressing 293T cell line [43]. Based on in vitro data and its high homology to ADAMTS2, ADAMTS3 had been thought to function in the proteolytic maturation of procollagens [61]. In mice, *Adamts3* deletion does not lead to collagen fibril assembly deficiencies but instead aborts lymphatic development [42]. In humans, mutations in *ADAMTS3* have similarly been shown to result in a lymphatic phenotype, while symptoms associated with procollagen cleavage defects are absent [56]. Although these and other publications have confirmed that both ADAMTS3 and CCBE1 are required for successful pro-VEGF-C activation, a direct interaction between VEGF-C and CCBE1 has never been demonstrated [41,62,63].

Both domains of CCBE1 accelerate the activation of VEGF-C independently [43,54,58], using different mechanisms. While the N-terminal domain of CCBE1 appears to facilitate pro-VEGF-C encounters with ADAMTS3, the C-terminal domain acts like a coenzyme [64]. From all VEGF-C-activating enzymes, only ADAMTS3 and PSA/KLK3 have been shown to be influenced by CCBE1.

6. Species-Specific Differences

Based on sequence similarity, in vitro substrate, and domain organization, ADAMTS2, -3, and -14 form the aminoprocollagen peptidase subgroup within the ADAMTS protein family. Species-specific differences in the function of these proteases are seen in vertebrates. In zebrafish, Adamts3 and Adamts14 compensate for each other, and only the double *Adamts3/Adamts14* knock-out shows a lymphatic phenotype comparable to the *Vegfc* knock-out [65]. Such compensation does not happen in Adamts3-deficient mice, which are completely devoid of functional lymphatics [42]. Whether the observation that human ADAMTS14 can activate VEGF-C in vitro [65] reflects species differences among mammals or whether it is an observation without a physiological equivalent is still unknown.

Important species differences have also been reported for the growth factors. In mice, VEGF-D is dispensable for the development of the lymphatic system [53], while this is not the case in zebrafish, where it is, e.g., required to form the medial and lateral facial lymphatics [66,67]. However, even murine VEGF-D reportedly differs from human VEGF-D in its inability to interact with mouse VEGFR-2 [68]. Exactly the opposite seems to be the case in zebrafish, where the VEGF-D-VEGFR-3 interaction was reported to be absent [69], implying that lymphangiogenesis might happen in zebrafish independently of

VEGFR-3. While direct demonstrations of the substrate specificities of the zebrafish Adamts proteases are still missing, it appears clear that zebrafish data are not easily extrapolated to mammals. Unfortunately, the same might be true for the extrapolation of mouse data to humans.

7. Which Cell Types Provide ADAMTS3 and CCBE1?

Since VEGF-C, ADAMTS3, and CCBE1 are all secreted proteins, immunohistochemistry cannot reveal their cellular origin. In the establishment of the early zebrafish lymphatics, *Pdgfra*-positive fibroblast populations appeared to be the source for *Vegfc*, *Adamts3*, *Adamts14*, and *Ccbe1*, as identified by single-cell RNA sequencing [65]. While in vitro data support the notion that fibroblasts are perhaps the dominating source for CCBE1 also in mammals [63], smooth muscle cells appear to make a significant contribution [14,70]. In some contexts, blood vascular endothelial cells appear to also be an important source of VEGF-C [71,72] and CCBE1 [73,74]. However, these are crude approximations of the actual cellular heterogeneity, and in non-homeostatic situations such as inflammation or cancer, other cell types, e.g., immune cells such as macrophages, are likely significant producers of both VEGF-C and VEGF-C-activating proteases [75–77].

8. Enigmatic Propeptides

The evolutionary origins of both propeptides of VEGF-C and VEGF-D are unclear. Unless assuming horizontal gene transfer, they have been conserved for hundreds of millions of years and can be found in virtually all invertebrate VEGF homologs [78–80]. Apart from the VEGFs, the only homologous sequences were found within larval silk proteins of the mosquito genus Chironomus [38,81], resulting in the nickname "silk homology domain" for the C-terminal propeptide.

Because disulfide bonds link the C- and the N-terminal propeptides of VEGF-C and VEGF-D, the first, constitutive cleavage by the protein convertase furin (or PC5 or PC7) does not remove any of the propeptides from VEGF-C or VEGF-D. Both propeptides are released simultaneously with the activating cleavage between the N-terminal propeptide and the VEGF homology domain (see Figures 3 and 4). With 80 and 192 amino acid residues, respectively, the N- and C-terminal propeptides of VEGF-C are significantly longer than typical propeptides. They also fold independently and are therefore also often referred to as N- and C-terminal *domains*. According to the current understanding, the propeptides serve multiple functions.

1. The heparin-binding C-terminal propeptide mediates ECM-association and cell surface (HSPG) binding [63,82].
2. Both propeptides collaborate in regulating receptor binding and activation [38,43]. While pro-VEGF-C binds VEGFR-3, it cannot (or only marginally) activate VEGFR-3. Thus, pro-VEGF-C and the individual VEGF-C propeptides, are competitive inhibitors of mature VEGF-C [43].
3. The presence of the C-terminal propeptide is required for efficient cleavage of the N-terminal propeptide by ADAMTS3 [63].

Analogous to VEGF-A, the heparin-binding properties are likely necessary for the correct spatio-temporal distribution of the growth factor and its activity. When the VEGF-C propeptides are grafted upon VEGF-A, the resulting blood vasculature was denser compared with VEGF-A-induced vasculature [83]. Vice versa, when the C-terminal domain of VEGF-C was replaced by the heparin-binding domain of VEGF-A, less but larger lymphatic vessels were generated, which localized preferentially to HSPG-rich structures such as basement membranes [84]. The heparin binding of VEGF-C is somewhat weaker compared to that of VEGF-A. Although most heparin-binding affinity resides in the C-terminal propeptide, mature VEGF-C is a heparin-binding growth factor. VEGF-A$_{165}$ binds tightest to heparin requiring 0.8 M NaCl for elution, while pro-VEGF-C and mature VEGF-C require elution concentrations of 0.435 and 0.265 M, respectively [82]. This might explain why both mature and pro-VEGF-C have a local effect and do not diffuse far [65]. That the C-terminal domain mediates the association or embedding of VEGF-C in the extracellular matrix was speculated shortly after its discovery [38] and was recently directly demonstrated in vitro [63]. Thus, pro-VEGF-C might be similar in this respect to latent TGF-β [85].

9. Changing Receptor Preferences with KLK3 and Cathepsin D

Based on N-terminal sequencing, two different mature, active forms were identified for VEGF-C and VEGF-D [38,86]. In the supernatant of 293 cells, the shorter mature form of VEGF-C was the dominant mature ("major") form, while, for VEGF-D, the longer mature form was dominant. While this indicated early on that different proteases are involved in VEGF-C and VEGF-D activation, it remained unknown which proteases were involved. In 2011, Leppänen et al. found that the shorter ("minor") form of active VEGF-D was not able to activate VEGFR-3 [87]. This finding was surprising as the activation of VEGFR-3 is considered a prerequisite of being lymphangiogenic. It also indicated for the first time that a lymphangiogenic growth factor can be converted into an angiogenic growth factor by proteolysis. At the same time, it explained why VEGF-D had been identified in some experimental settings as a powerful angiogenic growth factor [88]. While other research confirmed the disparity between VEGF-C and VEGF-D in terms of protease utilization for activation [41], the exact nature of the VEGF-D-activating proteases remained unknown until 2019 when Jha et al. tested whether their newly discovered VEGF-C-activating proteases PSA and Cathepsin D (CatD) could also activate VEGF-D [45]. In fact, CatD was able to generate the VEGFR-2-specific mature form of VEGF-D, which Leppänen et al. had described in 2011 [87]. Despite this, it remains to be shown which protease activates VEGF-D in vivo and whether there is a "physiological protease" equivalent to the VEGF-C-activating ADAMTS3. Perhaps VEGF-D is solely activated in non-homeostatic situations such as tissue damage. Nevertheless, also without any pathological challenge, VEGF-D knock-out mice display subtle alterations in some lymphatic networks [53,89]. These minor phenotypes could result from a lack of activated VEGF-D, but equally well from a lack of pro-VEGF-D (assuming it has some low level of activity) or possibly VEGF-C/VEGF-D heterodimers.

When comparing the effects of different VEGF-C- and VEGF-D-activating proteases [45], two trends are visible, which are summarized in Figure 5:

1. The shorter the N-terminus of the resulting mature growth factor, the lower its receptor binding affinity and receptor activation potential.
2. N-terminal shortening affects VEGF-C and VEGF-D very differently. While VEGF-C rapidly loses its potential to activate VEGFR-2 (through activation by ADAMTS3 or PSA), VEGF-D maintains much of its VEGFR-2 binding and activation potential. Vice versa, VEGF-D rapidly loses its VEGFR-3 binding and activation potential, whereas VEGF-C maintains much of it when processed to a similar degree.
3. Both VEGF-C and VEGF-D are completely inactivated with respect to their receptor tyrosine kinase activity by complete removal of their N-terminal helices, which, e.g., can be achieved by prolonged exposure to plasmin.

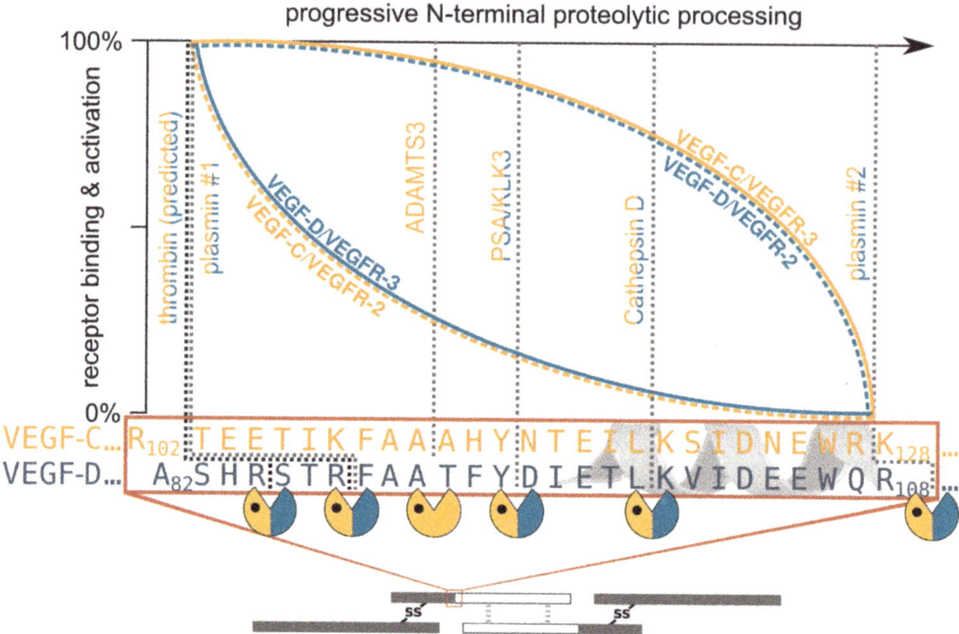

Figure 5. Biochemically, the group of lymphangiogenic activator enzymes is diverse. It includes a metalloproteinase (ADAMTS3), serine proteases (prostate-specific antigen (PSA)/KLK3, thrombin, and plasmin), and an aspartic protease (cathepsin D). VEGF-C and VEGF-D share all activating enzymes except for the most important one: ADAMTS3. ADAMTS3 is exclusive for VEGF-C and required for the physiologic activation of VEGF-C during developmental lymphangiogenesis [41–43]. VEGF-C and VEGF-D are differently affected by proteolytic processing. With progressing processing, VEGF-C largely maintains its lymphangiogenic properties but loses its angiogenic properties quickly. VEGF-D behaves precisely the opposite way: processing with Cathepsin D almost completely abolishes its lymphangiogenic properties and fully unmasks its angiogenic properties [45]. Extensive exposure of both VEGF-C and VEGF-D to plasmin abolishes all VEGFR-2 and VEGFR-3 binding properties.

10. Secondary Processing and Inactivation

At least in vitro, the longer forms of activated VEGF-C and VEGF-D can undergo additional cleavages, further shortening the N-terminus and modifying the receptor binding capabilities, e.g., CatD can remove the lymphangiogenic potential from plasmin-activated VEGF-D and the angiogenic potential from ADAMTS3-activated VEGF-C [45]. Finally, a cleavage by plasmin can inactivate VEGF-C and VEGF-D. However, such secondary (or tertiary) processing has not yet been demonstrated in vivo.

11. Other Cleavages

The activating, N-terminal cleavage of VEGF-D has also been proposed to be mediated by the protein convertases furin or PC5 [39]. While this is certainly a possibility, it seems unlikely that this represents a significant VEGF-D activation mechanism in vivo. The mature VEGF-D produced in furin-deficient Lovo cells upon transfection with furin could well be due to any other endogenous protease in Lovo cells. The requirement for furin in this system might occur if C-terminal furin processing was a prerequisite for the activating N-terminal cleavage. However, such a prerequisite seems not to exist for VEGF-C [41]. Vice versa, plasmin [44] or PSA [45] have not only been shown to perform the N-terminal processing of VEGF-D, but also the C-terminal processing. Nonetheless, this is also likely irrelevant since the protein convertases cleave constitutively inside the cell before VEGF-D ever has the chance to encounter plasmin or PSA.

12. Possible Involvement in Reproduction and Wound Healing

That VEGF-C is a possible substrate for kallikrein-like peptidases had been proposed before [90], but the identification of KLK3 (also known as prostate-specific antigen, PSA) was, nevertheless, surprising, especially because KLK3 is largely confined to sperm plasma. The presence of VEGF-C, CCBE1, and a VEGF-C-activating protease in sperm plasma [45] is too tempting not to speculate about a possible function of VEGF-C for reproductive biology, but such has not been confirmed yet. Mutations in KLK3 affect male fertility [91], but this is unsurprising since the main biological function of KLK3 is the degradation of gel-like seminogelins, which releases the sperm cells [92]. VEGF-A, which is also present in sperm plasma [93,94], had a modest effect on sperm motility [95]. Therefore, similar experiments were attempted with VEGF-C yielding very variable results (unpublished data by the author), perhaps due to the logistically challenging experimental setup.

Like KLK3, cathepsin D was also identified after an exhaustive analysis of bodily fluids for possible VEGF-C-cleaving activities [45]. VEGF-C is deposited by virtue of its C-terminal domain into the extracellular matrix and released by proteases [63]. Because VEGF-C accelerates wound healing [96,97], it appears possible that Cathepsin D provided by wound licking might activate latent ECM-embedded pro-VEGF-C. Thus, an instant angiogenic, lymphangiogenic, and immunologic stimulus would be provided. Compared to a single gene in humans, *KLK1* was several times duplicated in rodents, leading to at least 23 *KLK1* orthologs (some of which being pseudogenes), and some researchers believe that the evolutionary pressure to heal bite wounds rapidly and efficiently was driving this expansion [98].

13. Activating VEGF-C and VEGF-D in Cell Culture

While there are several cell lines that endogenously express VEGF-C or VEGF-D (most notably PC-3, from which VEGF-C was originally identified [99]), almost all experiments that require the expression of these growth factors have been performed by cDNA transfection. When the full-length wildtype cDNAs are used, the inactive pro-forms dominate in the cell culture supernatant of most cell lines (see Figure 3). Cells that express both CCBE1 and ADAMTS3 (such as cell lines derived from 293 cells) will process at least some of the pro-VEGF-C into mature, active VEGF-C. This endogenous background activation is sufficient to detect mature VEGF-C even in the absence of added proteases (see Figure 3). If these background activation bands are missing from a Western blot, the detection is likely not very sensitive, or something interferes with the physiological activation of VEGF-C by ADAMTS3. The degree of processing is relatively difficult to predict and appears to depend on cell density, stress level, cell culture medium, and—most importantly—VEGF-C expression levels. In any case, the processing is inefficient, and the 293T cell line that was used to generate the gel image in Figure 3 is among the cell lines that most efficiently activate VEGF-C endogenously.

14. Truncated cDNAs Are Used to Recombinantly Express Pre-Activated VEGF-C and VEGF-D

Therefore, when larger amounts of active VEGF-C are required, the solution has been to express a mutant VEGF-C cDNA, from which the sequences coding for the propeptides have been deleted ("ΔNΔC-VEGF-C"). All recombinant, commercially available VEGF-C and VEGF-D proteins are produced in this fashion. However, because the signal peptide's cleavage context is disturbed, the N-terminus of the resulting protein can differ from the endogenously activated VEGF-C. Only N-terminal sequencing can reveal which form of VEGF-C is present. With few exceptions (R&D Systems), vendors do not provide this information. The same is true for many scientific publications that use truncated cDNAs to express VEGF-C or VEGF-D. While it is possible to predict the signal peptidase's likely cleavage position, only N-terminal sequencing can give a definite answer. Many of the early experiments involving recombinant VEGF-D have used a truncated cDNA that results in a VEGF-D form, which is an intermediate between the VEGFR-2-monospecific and the VEGFR-2-/VEGFR-3-bispecific endogenous VEGF-D forms, making it difficult to interpret the data [68]. However, after the recent identification of the cleaving proteases, it became possible to generate specific mature forms by co-transfection of the protease with the full-length wildtype growth factor cDNA [43,45]. However, when using pre-activated forms of VEGF-C or VEGF-D, one should remember that it is unclear whether these exist as independent species in vivo. Pro-VEGF-C efficiently binds VEGFR-3 in the context of neuropilin-2, and the "in-situ" activation of pro-VEGF-C (while being bound to VEGFR-3) might be the standard mode of activation [43,63]. Interestingly, a transgenic mouse expressing pre-activated (ΔNΔC-VEGF-C) VEGF-C under the control of the keratin-14 promoter did not show the characteristic lymphatic phenotype in the skin as mice expressing VEGF-C from a full-length cDNA under the same promoter (unpublished data by the author) [100].

15. Modulation of Proteolytic Processing

Protease inhibitors have a veritable track record as drugs, targeting, e.g., viral proteases in AIDS and other viral infections [101], neutrophil elastase in lung diseases [102], and angiotensin-converting enzyme in cardiovascular diseases [103]. The opposite approach—promoting proteolysis—has also resulted in life-saving treatments, e.g., the use of tissue plasminogen activator (tPA) to dissolve blood clots in the immediate treatment of ischemic stroke [104].

Given the importance of the lymphatic system in many diseases [105], both VEGF-C and VEGF-D are likely worthwhile drug targets. Lymphedema, the swelling of organs or tissues due to an absent, hypoplastic, dysfunctional, or overloaded lymphatic network, represents a major clinical challenge because no causal, only symptomatic, treatments are available. The concept of pro-lymphangiogenic therapy to treat lymphedema has progressed to clinical trials using adenoviral VEGF-C gene therapy [106]. In these trials, pro-VEGF-C is produced from a full-length cDNA, relying on endogenous proteases for its activation. Using an adenovirus that produces pre-activated VEGF-C from a truncated cDNA would remove the requirement for endogenous proteases, which might or might not be a bottleneck. Alternatively, co-delivering CCBE1 and/or ADAMTS3 using an adenovirus cocktail could also boost the amount of active VEGF-C. However, the full-length cDNA of VEGF-C has been preferred over the truncated cDNA in most preclinical studies. Removing the N-terminal propeptide from VEGF-C results in an unpaired cysteine residue in VEGF-C's receptor-binding domain. This extra cysteine residue is conserved among all VEGF-C and VEGF-D orthologs but is absent from all other VEGF family members [99,107,108]. Therefore, when active VEGF-C or VEGF-D are expressed directly from a truncated cDNA, the monomeric growth factor can be the predominant species [103]. Monomeric growth factor exposes the dimerization interface to the environment and is predicted to interfere with receptor dimerization and activation. For crystallization studies, the extra cysteine residues can be mutated to promote dimerization [8,82], but the use of mutated proteins as biological drugs requires solid justification.

A continuous low-level supply with VEGF-C appears necessary to maintain the structure and functionality of heavily engaged lymphatic networks [70,109,110]. As an alternative to VEGF-C itself, a highly specific VEGF-C-activating protease might equally be suitable if it can activate endogenous ECM-embedded VEGF-C. Such VEGF-C activation might both act via stimulating lymphatic pumping [111] and by inducing a compensatory expansion of the lymphatic network [112].

In inflammatory and infectious diseases, the lymphatic network must manage the fluid balance during inflammatory swelling. Perhaps more importantly, there is increasing evidence that both innate and adaptive immunological responses are crucially dependent on the lymphatics during all stages of an immune response [113,114]. Thus, the activation of VEGF-C could be used as a generic means to boost any immune response like an adjuvant.

16. Proteolytic Activation of VEGF-C and VEGF-D in Cancer

The crucial role of tumor-associated lymphatics for metastasis was recognized early on [115–118], and VEGF-C/VEGF-D inhibition has been proposed to therapeutically block metastasis. Since the tumor-promoting effects of VEGF-C and VEGF-D likely require proteolytic processing, inhibition could not only target the growth factors or receptors, but also the activating proteases. In vitro, VEGF-C-expressing MCF-7 and MDA-MB-435 cells, which have been used for xenograft tumor models, are inefficient in activating the growth factor [115,117]. Therefore, it is assumed that the activating proteases are supplied in these xenograft models by the stromal tumor compartment, perhaps by fibroblasts, inflammatory or endothelial cells [119,120]. Harris et al. generated a mutated form of VEGF-D, which is resistant to proteolytic activation They showed that this mutant could not promote tumor growth and lymph node metastasis in a mouse tumor model [121].

Tumor cells can migrate and form distant metastases via two distinct pathways: via blood vessels (hematogenic spread) and via lymphatics (lymphogenic spread). VEGF-C and VEGF-D can stimulate both pathways. By stimulating lymphangiogenesis into the tumor periphery (and occasionally also into the tumor), these growth factors maximize the access of tumor cells to the lymphatic vasculature. VEGF-C further appears to actively prepare the downstream lymph nodes for arriving cancer cells [122]. If the tumor happens to express suitable proteases, it is likely that VEGF-C (and even more so VEGF-D) is activated into forms that mimic VEGF-A but which are not inhibited by current anti-angiogenic treatments [123]. Such angiogenic redundancy might be one of the reasons why VEGF-A treatment is much less universal as initially anticipated, and why in amenable cancers, initial treatment success is usually followed by the development of resistance [124].

However, anti-lymphangiogenic therapy is a double-edged sword [125] because tumor-associated lymphatics are crucially important for the immune response against the tumor. When VEGF-C action was blocked in a mouse tumor model treated with immunotherapy, the mice receiving the anti-VEGF-C treatment died earlier than those who did not receive the treatment [126]. Similarly, in a mouse glioblastoma model, VEGF-C could amplify the CD8+ T cell response against the tumor [127]. In order to be able to successfully target VEGF-C in cancer, a thorough understanding of the underlying molecular mechanisms is needed. This understanding might ultimately allow us to separate the metastasis-enhancing function of VEGF-C from the immune-response-enhancing function of VEGF-C. Such separation might involve activating VEGF-C into a largely VEGFR-3-specific form eliminating some (but not all) of VEGF-C's angiogenic features. However, since trafficking via the lymphatic neovasculature is integral to both tumor cell dissemination and immune response, separating these two functions appears unlikely to be feasible at the level of the VEGF-C/VEGFR-3 signaling axis. However, our understanding of the underlying molecular events is highly incomplete as we do not even know which specific proteases are activating VEGF-C in human cancers. A high-probability guess is that different proteases are involved depending on the cancer type.

17. Blocking VEGF-C and VEGF-D Activation

The proteolytic processing of VEGF-C and VEGF-D has been so far experimentally blocked only by mutagenesis of the cleavage sites. Furin and related protein convertases cleave VEGF-C after the double arginines (R226,227). Joukov et al. reported that mutating these arginines into serines (R226,227S) mostly blocked VEGF-C processing [38]. This is surprising since the N-terminal cleavage site should have been still subject to proteolytic attack because the first, constitutive C-terminal cleavage and the second, N-terminal cleavage were subsequently shown to occur independently of each other [41]. To generate an even more activation-resistant form of VEGF-D, Harris et al. mutated, in addition to the C-terminal cleavage site, the major N-terminal cleavage site and reported, similar to Joukov et al., almost complete abrogation of VEGF-D activation [121]. It is unclear why the unmutated minor N-terminal cleavage site in this protein did not result at least in partial activation. After all, in the same 293 EBNA cell line, the minor N-terminal cleavage site had been shown to account for approximately 20% of the activated protein [86]. However, a therapeutic effect might not require full inhibition of cleavage because pro-VEGF-C acts as an antagonist of mature VEGF-C, and therefore, a low level of cleavage might be acceptable [43].

18. Lymphedema and Genetic Lesions Affecting the Activation of VEGF-C

Lymphedema is traditionally categorized into either primary or secondary lymphedema. Most lymphedema cases fall into the secondary category, resulting from various external insults to the lymphatic system, surgery and infection being most common [128]. On the other hand, primary lymphedema results from genetic lesions, which can be inherited or acquired during development. The latter makes them more challenging to identify since the genetic lesion might be present only in a subset of cells or organs [129]. For roughly 40% of primary lymphedema cases, the underlying genetic lesion can be identified, and a clinical guide for this process has been established [130]. However, secondary lymphedema also has a genetic component [131–133], and thus, primary and secondary lymphedema should be considered the endpoints of a continuous spectrum.

All mutations that disrupt the VEGF-C/VEGFR-3 signaling pathway result in hereditary lymphedema (Figure 6). While the most common type of hereditary lymphedema is caused by a mutation in the VEGF-C receptor [134,135], any signaling pathway components can be affected, including the proteolytic activation of VEGF-C. Thus, in human lymphedema patients, disease-causing mutations have been found in VEGF-C itself [136,137], in its activating protease ADAMTS3 [62,63], and the cofactor CCBE1 [57,138]. Genetic lesions in *FAT4* result in a phenotype closely resembling the phenotypes caused by mutations in *CCBE1* or *ADAMTS3*. Only recently, Hennekam Syndrome was split into three different subtypes depending on the underlying genetic lesion (Hennekam Syndrome Type 1, 2, and 3).

Interestingly, the primary function of FAT4 is likely unrelated to VEGF-C processing or VEGFR-3 signaling. Instead, it appears necessary for the flow-dependent establishment of lymphatic endothelial cell polarity [139]. However, in vitro analysis of FAT4 has been hampered because it is a very large protein with a highly repetitive structure [140]. Besides VEGF-C/VEGFR-3 signaling pathway genes known to be compromised in hereditary lymphedema or Hennekam Syndrome, Table 1 also lists selected other genes known to be causative for hereditary diseases featuring lymphedema as a cardinal symptom. Furthermore, Table 1 contains, in addition to ADAMTS3, the genes of all other proteases reportedly able to activate VEGF-C and VEGF-D. For all of these, genetic lesions have been described, but lymphedema has not been associated with any of them, arguing that they are not required for the development of the lymphatic system.

Table 1. Important genes for lymphatic development and/or VEGF-C activation, which are involved in human hereditary disorders. Mutations in the VEGF-C-activating proteases are not associated with any lymphatic phenotype except for ADAMTS3, arguing that only ADAMTS3 is essential for lymphatic development.

Gene	Protein	Human Disease (OMIM)	Remarks
Genes within the VEGF-C/VEGFR-3 signaling pathway; for clinical details see [130], and for molecular details [141]			
VEGFC	Vascular Endothelial Growth Factor-C (VEGF-C)	Hereditary lymphedema type 1D (615907)	VEGF-C is the primary growth factor for lymphatic endothelial cells.
FLT4	Vascular Endothelial Growth Factor Receptor-3 (VEGFR-3)	Hereditary lymphedema type 1A (Milroy disease, 153100)	VEGFR-3 is the primary receptor of VEGF-C.
CCBE1	Collagen and calcium-binding EGF domain-containing protein 1	Hennekam lymphangiectasia-lymphedema syndrome type 1 (235510)	Enhances the processing of VEGF-C by ADAMTS3 and KLK3.
ADAMTS3	A disintegrin and metalloproteinase with thrombospondin motifs 3	Hennekam lymphangiectasia-lymphedema syndrome type 3 (618154)	ADAMTS3 catalyzes the final step in the activation of VEGF-C.

Table 1. *Cont.*

Gene	Protein	Human Disease (OMIM)	Remarks
Genes coding for proteases that can activate VEGF-C and/or VEGF-D (no lymphatic phenotype reported)			
ADAMTS14	A disintegrin and metalloproteinase with thrombospondin motifs 14	Association with age of onset in tendinopathy [142]	ADAMTS14 can activate VEGF-C in vitro [65].
KLK3	Kallikrein-like peptidase 3, Prostate-specific antigen (PSA)	Association with human fertility [91]	KLK3/PSA is most commonly known as a prostate cancer marker [143].
CTSD	Cathepsin D	Neuronal Ceroid Lipofuscinosis type 10 (610127)	CTSD deficiency causes a neurodegenerative disorder [144].
F2	Prothrombin/Thrombin	Hereditary thrombophilia type 1 (188050)	Specific F2 mutations increase the risk of venous thromboembolism [145]. Thrombin potentiates vascular endothelial growth factor- (VEGF-) induced endothelial cell proliferation [146].
PLG	Plasminogen/Plasmin	Plasminogen deficiency type 1 (217090)	PLG deficiency leads to pathological fibrin deposition but no increased risk of thrombosis [147].
Selected other "lymphedema" genes; for a comprehensive listing, see [130]			
FOXC2	Forkhead box protein C2	Lymphedema distichiasis syndrome (153400)	The maturation of lymphatic vessels and the formation of lymphatic valves requires FOXC2 [148,149].
GJC2	Connexin 47	Hereditary lymphedema type 1C (613480)	CJC2 is a gap junction protein that enables communication between lymphatic endothelial cells [150].
FAT4	Protocadherin Fat 4	Hennekam lymphangiectasia-lymphedema syndrome type 2 (616006)	FAT4 is required for lymphatic endothelial cell polarity and might influence VEGFR-3 signaling [139].
		Van Maldergem syndrome type 2 (615546)	Van Maldergem syndrome 2 has overlapping features with Hennekam syndrome type 2 but none or only infrequent lymphatic involvement [151].

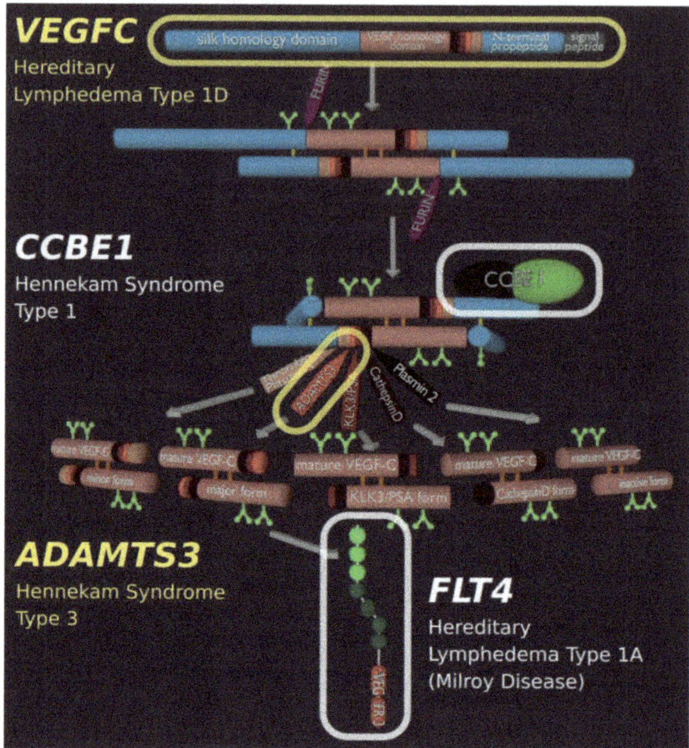

Figure 6. Genes of the VEGF-C/VEGFR-3 signaling pathway known to be involved in hereditary lymphedema. Although different components of the signaling pathway can be affected, the largest fraction of cases are caused by mutations in the *FLT4* gene, which codes for the VEGF-C receptor VEGFR-3.

19. Outlook: Molecular Nudging

With the first successes in Crispr-Cas clinical trials, genetic deficiencies within the VEGF-C/VEGFR-3 signaling pathways appear at least theoretically amenable for repair. However, even cutting-edge trials limit themselves at the moment to cells that can be easily modified ex vivo (blood diseases such as sickle cell disease and β-thalassemia) [152] or to very localized targets [153]. We are still far from a systemic repair of solid tissues, which would be needed since the lymphatic system penetrates almost all our bodies' organs. Since at least a fraction of the VEGF-C appears to originate from blood vascular endothelial cells, a vascular-targeted repair appears possible [154]. If sufficiently specific, the systemic delivery of regulatory factors such as CCBE1 or ADAMTS3 might alternatively result in a widespread low-level activation ("molecular nudging") of endogenous VEGF-C and a therapeutic effect. While such interventions do not reverse developmental routes already taken, they still might significantly improve life quality.

For cancer, being the prototype of a moving drug target, molecular nudging is not likely to have any impact. While a multitargeted anti-VEGF-A/-C/-D therapy might result in improved survival, any progress in this area will likely be incremental since using alternative tumor angiogenesis factors is only one of many escape mechanisms that tumors can deploy [124].

Author Contributions: Writing—original draft preparation, M.J.; writing—review and editing, J.K., H.B., M.J.; visualization, H.B., M.J.; supervision, project administration, and funding acquisition, M.J. All authors have read and agreed to the published version of the manuscript.

Funding: This research was funded by the Päivikki and Sakari Sohlberg Foundation, the Novo Nordisk Foundation (#21036), and the Academy of Finland (#337430, #337120). M.J. was supported by the Paulo Foundation and the Einar and Karin Stroem Foundation for Medical Research; H.B. was supported by a Fellowship from the Finnish National Agency for Education (TM-20-11320).

Institutional Review Board Statement: Not applicable.

Informed Consent Statement: Not applicable.

Data Availability Statement: Not applicable.

Conflicts of Interest: The authors declare no conflict of interest.

References

1. de Groof, A.; Guelen, L.; Deijs, M.; van der Wal, Y.; Miyata, M.; Ng, K.S.; van Grinsven, L.; Simmelink, B.; Biermann, Y.; Grisez, L.; et al. A Novel Virus Causes Scale Drop Disease in Lates Calcarifer. *PLOS Pathog.* **2015**, *11*, e1005074. [CrossRef]
2. Lyttle, D.J.; Fraser, K.M.; Fleming, S.B.; Mercer, A.A.; Robinson, A.J. Homologs of Vascular Endothelial Growth Factor Are Encoded by the Poxvirus Orf Virus. *J. Virol.* **1994**, *68*, 84–92. [CrossRef]
3. Meyer, M.; Clauss, M.; Lepple-Wienhues, A.; Waltenberger, J.; Augustin, H.G.; Ziche, M.; Lanz, C.; Büttner, M.; Rziha, H.-J.; Dehio, C. A Novel Vascular Endothelial Growth Factor Encoded by Orf Virus, VEGF-E, Mediates Angiogenesis via Signalling through VEGFR-2 (KDR) but Not VEGFR-1 (Flt-1) Receptor Tyrosine Kinases. *EMBO J.* **1999**, *18*, 363–374. [CrossRef] [PubMed]
4. Ogawa, S.; Oku, A.; Sawano, A.; Yamaguchi, S.; Yazaki, Y.; Shibuya, M. A Novel Type of Vascular Endothelial Growth Factor, VEGF-E (NZ-7 VEGF), Preferentially Utilizes KDR/Flk-1 Receptor and Carries a Potent Mitotic Activity without Heparin-Binding Domain. *J. Biol. Chem.* **1998**, *273*, 31273–31282. [CrossRef]
5. Yamazaki, Y.; Matsunaga, Y.; Tokunaga, Y.; Obayashi, S.; Saito, M.; Morita, T. Snake Venom Vascular Endothelial Growth Factors (VEGF-Fs) Exclusively Vary Their Structures and Functions among Species. *J. Biol. Chem.* **2009**, *284*, 9885–9891. [CrossRef] [PubMed]
6. Wiesmann, C.; Fuh, G.; Christinger, H.W.; Eigenbrot, C.; Wells, J.A.; de Vos, A.M. Crystal Structure at 1.7 Å Resolution of VEGF in Complex with Domain 2 of the Flt-1 Receptor. *Cell* **1997**, *91*, 695–704. [CrossRef]
7. Fuh, G.; Li, B.; Crowley, C.; Cunningham, B.; Wells, J.A. Requirements for Binding and Signaling of the Kinase Domain Receptor for Vascular Endothelial Growth Factor. *J. Biol. Chem.* **1998**, *273*, 11197–11204. [CrossRef]
8. Leppänen, V.-M.; Prota, A.E.; Jeltsch, M.; Anisimov, A.; Kalkkinen, N.; Strandin, T.; Lankinen, H.; Goldman, A.; Ballmer-Hofer, K.; Alitalo, K. Structural Determinants of Growth Factor Binding and Specificity by VEGF Receptor 2. *Proc. Natl. Acad. Sci. USA* **2010**, *107*, 2425–2430. [CrossRef] [PubMed]
9. Leppänen, V.-M.; Tvorogov, D.; Kisko, K.; Prota, A.E.; Jeltsch, M.; Anisimov, A.; Markovic-Mueller, S.; Stuttfeld, E.; Goldie, K.N.; Ballmer-Hofer, K.; et al. Structural and Mechanistic Insights into VEGF Receptor 3 Ligand Binding and Activation. *Proc. Natl. Acad. Sci. USA* **2013**, *110*, 12960–12965. [CrossRef]
10. Apte, R.S.; Chen, D.S.; Ferrara, N. VEGF in Signaling and Disease: Beyond Discovery and Development. *Cell* **2019**, *176*, 1248–1264. [CrossRef]
11. Simons, M.; Gordon, E.; Claesson-Welsh, L. Mechanisms and Regulation of Endothelial VEGF Receptor Signalling. *Nat. Rev. Mol. Cell Biol.* **2016**, *17*, 611–625. [CrossRef] [PubMed]
12. Carmeliet, P.; Ferreira, V.; Breier, G.; Pollefeyt, S.; Kieckens, L.; Gertsenstein, M.; Fahrig, M.; Vandenhoeck, A.; Harpal, K.; Eberhardt, C.; et al. Abnormal Blood Vessel Development and Lethality in Embryos Lacking a Single VEGF Allele. *Nature* **1996**, *380*, 435–439. [CrossRef] [PubMed]
13. Ferrara, N.; Carver-Moore, K.; Chen, H.; Dowd, M.; Lu, L.; O'Shea, K.S.; Powell-Braxton, L.; Hillan, K.J.; Moore, M.W. Heterozygous Embryonic Lethality Induced by Targeted Inactivation of the VEGF Gene. *Nature* **1996**, *380*, 439–442. [CrossRef]
14. Karkkainen, M.J.; Haiko, P.; Sainio, K.; Partanen, J.; Taipale, J.; Petrova, T.V.; Jeltsch, M.; Jackson, D.G.; Talikka, M.; Rauvala, H.; et al. Vascular Endothelial Growth Factor C Is Required for Sprouting of the First Lymphatic Vessels from Embryonic Veins. *Nat. Immunol.* **2003**, *5*, 74–80. [CrossRef]
15. Rodamilans, B.; Shan, H.; Pasin, F.; García, J.A. Plant Viral Proteases: Beyond the Role of Peptide Cutters. *Front. Plant Sci.* **2018**, *9*. [CrossRef]
16. van der Hoorn, R.A.L.; Rivas, S. Unravelling the Mode of Action of Plant Proteases. *New Phytol.* **2018**, *218*, 879–881. [CrossRef]
17. Künnapuu, J.; Björkgren, I.; Shimmi, O. The Drosophila DPP Signal Is Produced by Cleavage of Its Proprotein at Evolutionary Diversified Furin-Recognition Sites. *Proc. Natl. Acad. Sci. USA* **2009**, *106*, 8501–8506. [CrossRef]
18. Turk, B.; Turk, D.; Turk, V. Protease Signalling: The Cutting Edge. *EMBO J.* **2012**, *31*, 1630–1643. [CrossRef] [PubMed]
19. Vempati, P.; Popel, A.S.; Mac Gabhann, F. Extracellular Regulation of VEGF: Isoforms, Proteolysis, and Vascular Patterning. *Cytokine Growth Factor Rev.* **2014**, *25*, 1–19. [CrossRef]
20. Bridgett, S.; Dellett, M.; Simpson, D.A. RNA-Sequencing Data Supports the Existence of Novel VEGFA Splicing Events but Not of VEGFA Xxx b Isoforms. *Sci. Rep.* **2017**, *7*, 58. [CrossRef]

21. Keyt, B.A.; Berleau, L.T.; Nguyen, H.V.; Chen, H.; Heinsohn, H.; Vandlen, R.; Ferrara, N. The Carboxyl-Terminal Domain(111165) of Vascular Endothelial Growth Factor Is Critical for Its Mitogenic Potency. *J. Biol. Chem.* **1996**, *271*, 7788–7795. [CrossRef] [PubMed]

22. Park, J.E.; Keller, G.A.; Ferrara, N. The Vascular Endothelial Growth Factor (VEGF) Isoforms: Differential Deposition into the Subepithelial Extracellular Matrix and Bioactivity of Extracellular Matrix-Bound VEGF. *Mol. Biol. Cell* **1993**, *4*, 1317–1326. [CrossRef]

23. Carmeliet, P.; Ng, Y.-S.; Nuyens, D.; Theilmeier, G.; Brusselmans, K.; Cornelissen, I.; Ehler, E.; Kakkar, V.V.; Stalmans, I.; Mattot, V.; et al. Impaired Myocardial Angiogenesis and Ischemic Cardiomyopathy in Mice Lacking the Vascular Endothelial Growth Factor Isoforms VEGF 164 and VEGF 188. *Nat. Med.* **1999**, *5*, 495–502. [CrossRef]

24. Ruhrberg, C.; Gerhardt, H.; Golding, M.; Watson, R.; Ioannidou, S.; Fujisawa, H.; Betsholtz, C.; Shima, D.T. Spatially Restricted Patterning Cues Provided by Heparin-Binding VEGF-A Control Blood Vessel Branching Morphogenesis. *Genes Dev.* **2002**, *16*, 2684–2698. [CrossRef]

25. Gerhardt, H.; Golding, M.; Fruttiger, M.; Ruhrberg, C.; Lundkvist, A.; Abramsson, A.; Jeltsch, M.; Mitchell, C.; Alitalo, K.; Shima, D.; et al. VEGF Guides Angiogenic Sprouting Utilizing Endothelial Tip Cell Filopodia. *J. Cell Biol.* **2003**, *161*, 1163–1177. [CrossRef] [PubMed]

26. Plouët, J.; Moro, F.; Bertagnolli, S.; Coldeboeuf, N.; Mazarguil, H.; Clamens, S.; Bayard, F. Extracellular Cleavage of the Vascular Endothelial Growth Factor 189-Amino Acid Form by Urokinase Is Required for Its Mitogenic Effect. *J. Biol. Chem.* **1997**, *272*, 13390–13396. [CrossRef]

27. Houck, K.A.; Leung, D.W.; Rowland, A.M.; Winer, J.; Ferrara, N. Dual Regulation of Vascular Endothelial Growth Factor Bioavailability by Genetic and Proteolytic Mechanisms. *J. Biol. Chem.* **1992**, *267*, 26031–26037. [CrossRef]

28. Ferrara, N. Binding to the Extracellular Matrix and Proteolytic Processing: Two Key Mechanisms Regulating Vascular Endothelial Growth Factor Action. *Mol. Biol. Cell* **2010**, *21*, 687–690. [CrossRef] [PubMed]

29. Uslu, Ö.; Herold, J.; Kanse, S.M. VEGF-A-Cleavage by FSAP and Inhibition of Neo-Vascularization. *Cells* **2019**, *8*, 1396. [CrossRef]

30. Lee, S.; Jilani, S.M.; Nikolova, G.V.; Carpizo, D.; Iruela-Arispe, M.L. Processing of VEGF-A by Matrix Metalloproteinases Regulates Bioavailability and Vascular Patterning in Tumors. *J. Cell Biol.* **2005**, *169*, 681–691. [CrossRef]

31. Eming, S.A.; Krieg, T. Molecular Mechanisms of VEGF-A Action during Tissue Repair. *J. Investig. Dermatol. Symp. Proc.* **2006**, *11*, 79–86. [CrossRef] [PubMed]

32. Hoffmann, D.C.; Willenborg, S.; Koch, M.; Zwolanek, D.; Müller, S.; Becker, A.-K.A.; Metzger, S.; Ehrbar, M.; Kurschat, P.; Hellmich, M.; et al. Proteolytic Processing Regulates Placental Growth Factor Activities. *J. Biol. Chem.* **2013**, *288*, 17976–17989. [CrossRef]

33. Makinen, T.; Olofsson, B.; Karpanen, T.; Hellman, U.; Soker, S.; Klagsbrun, M.; Eriksson, U.; Alitalo, K. Differential Binding of Vascular Endothelial Growth Factor B Splice and Proteolytic Isoforms to Neuropilin-1. *J. Biol. Chem.* **1999**, *274*, 21217–21222. [CrossRef] [PubMed]

34. Olofsson, B.; Pajusola, K.; von Euler, G.; Chilov, D.; Alitalo, K.; Eriksson, U. Genomic Organization of the Mouse and Human Genes for Vascular Endothelial Growth Factor B (VEGF-B) and Characterization of a Second Splice Isoform. *J. Biol. Chem.* **1996**, *271*, 19310–19317. [CrossRef] [PubMed]

35. Rawlings, N.D.; Barrett, A.J.; Thomas, P.D.; Huang, X.; Bateman, A.; Finn, R.D. The MEROPS Database of Proteolytic Enzymes, Their Substrates and Inhibitors in 2017 and a Comparison with Peptidases in the PANTHER Database. *Nucleic Acids Res.* **2018**, *46*, D624–D632. [CrossRef]

36. Yamazaki, Y.; Morita, T. Molecular and Functional Diversity of Vascular Endothelial Growth Factors. *Mol. Divers.* **2006**, *10*, 515–527. [CrossRef]

37. Stenberg, L.M.; Brown, M.A.; Nilsson, E.; Ljungberg, O.; Stenflo, J. A Functional Prothrombin Gene Product Is Synthesized by Human Kidney Cells. *Biochem. Biophys. Res. Commun.* **2001**, *280*, 1036–1041. [CrossRef] [PubMed]

38. Joukov, V.; Sorsa, T.; Kumar, V.; Jeltsch, M.; Claesson-Welsh, L.; Cao, Y.; Saksela, O.; Kalkkinen, N.; Alitalo, K. Proteolytic Processing Regulates Receptor Specificity and Activity of VEGF-C. *EMBO J.* **1997**, *16*, 3898–3911. [CrossRef]

39. McColl, B.K.; Paavonen, K.; Karnezis, T.; Harris, N.C.; Davydova, N.; Rothacker, J.; Nice, E.C.; Harder, K.W.; Roufail, S.; Hibbs, M.L.; et al. Proprotein Convertases Promote Processing of VEGF-D, a Critical Step for Binding the Angiogenic Receptor VEGFR-2. *FASEB J.* **2007**, *21*, 1088–1098. [CrossRef] [PubMed]

40. Siegfried, G.; Basak, A.; Cromlish, J.A.; Benjannet, S.; Marcinkiewicz, J.; Chrétien, M.; Seidah, N.G.; Khatib, A.-M. The Secretory Proprotein Convertases Furin, PC5, and PC7 Activate VEGF-C to Induce Tumorigenesis. *J. Clin. Investig.* **2003**, *111*, 1723–1732. [CrossRef]

41. Bui, H.M.; Enis, D.; Robciuc, M.R.; Nurmi, H.J.; Cohen, J.; Chen, M.; Yang, Y.; Dhillon, V.; Johnson, K.; Zhang, H.; et al. Proteolytic Activation Defines Distinct Lymphangiogenic Mechanisms for VEGFC and VEGFD. *J. Clin. Investig.* **2016**, *126*, 2167–2180. [CrossRef]

42. Janssen, L.; Dupont, L.; Bekhouche, M.; Noel, A.; Leduc, C.; Voz, M.; Peers, B.; Cataldo, D.; Apte, S.S.; Dubail, J.; et al. ADAMTS3 Activity Is Mandatory for Embryonic Lymphangiogenesis and Regulates Placental Angiogenesis. *Angiogenesis* **2015**, 1–13. [CrossRef] [PubMed]

43. Jeltsch, M.; Jha, S.K.; Tvorogov, D.; Anisimov, A.; Leppänen, V.-M.; Holopainen, T.; Kivelä, R.; Ortega, S.; Kärpanen, T.; Alitalo, K. CCBE1 Enhances Lymphangiogenesis via A Disintegrin and Metalloprotease With Thrombospondin Motifs-3–Mediated Vascular Endothelial Growth Factor-C Activation. *Circulation* **2014**, *129*, 1962–1971. [CrossRef]

44. McColl, B.K.; Baldwin, M.E.; Roufail, S.; Freeman, C.; Moritz, R.L.; Simpson, R.J.; Alitalo, K.; Stacker, S.A.; Achen, M.G. Plasmin Activates the Lymphangiogenic Growth Factors VEGF-C and VEGF-D. *J. Exp. Med.* **2003**, *198*, 863–868. [CrossRef]

45. Jha, S.K.; Rauniyar, K.; Chronowska, E.; Mattonet, K.; Maina, E.W.; Koistinen, H.; Stenman, U.-H.; Alitalo, K.; Jeltsch, M. KLK3/PSA and Cathepsin D Activate VEGF-C and VEGF-D. *eLife* **2019**, *8*, e44478. [CrossRef] [PubMed]

46. Lim, L.; Bui, H.; Farrelly, O.; Yang, J.; Li, L.; Enis, D.; Ma, W.; Chen, M.; Oliver, G.; Welsh, J.D.; et al. Hemostasis Stimulates Lymphangiogenesis through Release and Activation of VEGFC. *Blood* **2019**, *134*, 1764–1775. [CrossRef]

47. Kärpänen, T.; Heckman, C.A.; Keskitalo, S.; Jeltsch, M.; Ollila, H.; Neufeld, G.; Tamagnone, L.; Alitalo, K. Functional Interaction of VEGF-C and VEGF-D with Neuropilin Receptors. *FASEB J.* **2006**, *20*, 1462–1472. [CrossRef]

48. Wartiovaara, U.; Salven, P.; Mikkola, H.; Lassila, R.; Kaukonen, J.; Joukov, V.; Orpana, A.; Ristimäki, A.; Heikinheimo, M.; Joensuu, H.; et al. Peripheral Blood Platelets Express VEGF-C and VEGF Which Are Released during Platelet Activation. *Thromb. Haemost.* **1998**, *80*, 171–175. [CrossRef] [PubMed]

49. Akgül, A.; Cirak, M.; Birinci, T. Applications of Platelet-Rich Plasma in Lymphedema. *Lymphat. Res. Biol.* **2016**, *14*, 206–209. [CrossRef] [PubMed]

50. Margolis, D.J.; Kantor, J.; Santanna, J.; Strom, B.L.; Berlin, J.A. Effectiveness of Platelet Releasate for the Treatment of Diabetic Neuropathic Foot Ulcers. *Diabetes Care* **2001**, *24*, 483–488. [CrossRef] [PubMed]

51. Lee, H.; Chae, S.; Park, J.; Bae, J.; Go, E.-B.; Kim, S.-J.; Kim, H.; Hwang, D.; Lee, S.-W.; Lee, S.-Y. Comprehensive Proteome Profiling of Platelet Identified a Protein Profile Predictive of Responses to An Antiplatelet Agent Sarpogrelate. *Mol. Cell. Proteom.* **2016**, *15*, 3461–3472. [CrossRef]

52. Yadav, D.B.; Reyes, A.E.; Gupta, P.; Vernes, J.-M.; Meng, Y.G.; Schweiger, M.G.; Stainton, S.L.; Fuh, G.; Fielder, P.J.; Kamath, A.V.; et al. Complex Formation of Anti-VEGF-C with VEGF-C Released during Blood Coagulation Resulted in an Artifact in Its Serum Pharmacokinetics. *Pharmacol. Res. Perspect.* **2020**, *8*, e00573. [CrossRef]

53. Baldwin, M.E.; Halford, M.M.; Roufail, S.; Williams, R.A.; Hibbs, M.L.; Grail, D.; Kubo, H.; Stacker, S.A.; Achen, M.G. Vascular Endothelial Growth Factor D Is Dispensable for Development of the Lymphatic System. *Mol. Cell. Biol.* **2005**, *25*, 2441–2449. [CrossRef] [PubMed]

54. Lee, J.; Gray, A.; Yuan, J.; Luoh, S.M.; Avraham, H.; Wood, W.I. Vascular Endothelial Growth Factor-Related Protein: A Ligand and Specific Activator of the Tyrosine Kinase Receptor Flt4. *Proc. Natl. Acad. Sci. USA* **1996**, *93*, 1988–1992. [CrossRef]

55. Bugge, T.H.; Flick, M.J.; Daugherty, C.C.; Degen, J.L. Plasminogen Deficiency Causes Severe Thrombosis but Is Compatible with Development and Reproduction. *Genes Dev.* **1995**, *9*, 794–807. [CrossRef]

56. Kikuchi, S.; Yamanouchi, Y.; Li, L.; Kobayashi, K.; Ijima, H.; Miyazaki, R.; Tsuchiya, S.; Hamaguchi, H. Plasminogen with Type-I Mutation Is Polymorphic in the Japanese Population. *Hum. Genet.* **1992**, *90*, 7–11. [CrossRef] [PubMed]

57. Alders, M.; Hogan, B.M.; Gjini, E.; Salehi, F.; Al-Gazali, L.; Hennekam, E.A.; Holmberg, E.E.; Mannens, M.M.A.M.; Mulder, M.F.; Offerhaus, G.J.A.; et al. Mutations in CCBE1 Cause Generalized Lymph Vessel Dysplasia in Humans. *Nat. Genet.* **2009**, *41*, 1272–1274. [CrossRef]

58. Hennekam, R.C.M.; Geerdink, R.A.; Hamel, B.C.J.; Hennekam, F.A.M.; Kraus, P.; Rammeloo, J.A.; Tillemans, A.A.W. Autosomal Recessive Intestinal Lymphangiectasia and Lymphedema, with Facial Anomalies and Mental Retardation. *Am. J. Med. Genet.* **1989**, *34*, 593–600. [CrossRef] [PubMed]

59. Hogan, B.M.; Bos, F.L.; Bussmann, J.; Witte, M.; Chi, N.C.; Duckers, H.J.; Schulte-Merker, S. Ccbe1 Is Required for Embryonic Lymphangiogenesis and Venous Sprouting. *Nat. Genet.* **2009**, *41*, 396–398. [CrossRef]

60. Bos, F.L.; Caunt, M.; Peterson-Maduro, J.; Planas-Paz, L.; Kowalski, J.; Karpanen, T.; Van Impel, A.; Tong, R.; Ernst, J.A.; Korving, J.; et al. CCBE1 Is Essential for Mammalian Lymphatic Vascular Development and Enhances the Lymphangiogenic Effect of Vascular Endothelial Growth Factor-C In Vivo. *Circ. Res.* **2011**, *109*, 486–491. [CrossRef]

61. Fernandes, R.J.; Hirohata, S.; Engle, J.M.; Colige, A.; Cohn, D.H.; Eyre, D.R.; Apte, S.S. Procollagen II Amino Propeptide Processing by ADAMTS-3 INSIGHTS ON DERMATOSPARAXIS. *J. Biol. Chem.* **2001**, *276*, 31502–31509. [CrossRef] [PubMed]

62. Brouillard, P.; Dupont, L.; Helaers, R.; Coulie, R.; Tiller, G.E.; Peeden, J.; Colige, A.; Vikkula, M. Loss of ADAMTS3 Activity Causes Hennekam Lymphangiectasia–Lymphedema Syndrome 3. *Hum. Mol. Genet.* **2017**, *26*, 4095–4104. [CrossRef]

63. Jha, S.K.; Rauniyar, K.; Karpanen, T.; Leppänen, V.-M.; Brouillard, P.; Vikkula, M.; Alitalo, K.; Jeltsch, M. Efficient Activation of the Lymphangiogenic Growth Factor VEGF-C Requires the C-Terminal Domain of VEGF-C and the N-Terminal Domain of CCBE1. *Sci. Rep.* **2017**, *7*, 4916. [CrossRef] [PubMed]

64. Le Guen, L.; Karpanen, T.; Schulte, D.; Harris, N.C.; Koltowska, K.; Roukens, G.; Bower, N.I.; van Impel, A.; Stacker, S.A.; Achen, M.G.; et al. Ccbe1 Regulates Vegfc-Mediated Induction of Vegfr3 Signaling during Embryonic Lymphangiogenesis. *Development* **2014**, *141*, 1239–1249. [CrossRef]

65. Wang, G.; Muhl, L.; Padberg, Y.; Dupont, L.; Peterson-Maduro, J.; Stehling, M.; le Noble, F.; Colige, A.; Betsholtz, C.; Schulte-Merker, S.; et al. Specific Fibroblast Subpopulations and Neuronal Structures Provide Local Sources of Vegfc-Processing Components during Zebrafish Lymphangiogenesis. *Nat. Commun.* **2020**, *11*, 2724. [CrossRef] [PubMed]

66. Bower, N.I.; Vogrin, A.J.; Guen, L.L.; Chen, H.; Stacker, S.A.; Achen, M.G.; Hogan, B.M. Vegfd Modulates Both Angiogenesis and Lymphangiogenesis during Zebrafish Embryonic Development. *Development* **2017**, *144*, 507–518. [CrossRef]

67. Astin, J.W.; Haggerty, M.J.L.; Okuda, K.S.; Le Guen, L.; Misa, J.P.; Tromp, A.; Hogan, B.M.; Crosier, K.E.; Crosier, P.S. Vegfd Can Compensate for Loss of Vegfc in Zebrafish Facial Lymphatic Sprouting. *Development* **2014**, *141*, 2680–2690. [CrossRef] [PubMed]
68. Baldwin, M.E.; Catimel, B.; Nice, E.C.; Roufail, S.; Hall, N.E.; Stenvers, K.L.; Karkkainen, M.J.; Alitalo, K.; Stacker, S.A.; Achen, M.G. The Specificity of Receptor Binding by Vascular Endothelial Growth Factor-D Is Different in Mouse and Man. *J. Biol. Chem.* **2001**, *276*, 19166–19171. [CrossRef] [PubMed]
69. Vogrin, A.J.; Bower, N.I.; Gunzburg, M.J.; Roufail, S.; Okuda, K.S.; Paterson, S.; Headey, S.J.; Stacker, S.A.; Hogan, B.M.; Achen, M.G. Evolutionary Differences in the Vegf/Vegfr Code Reveal Organotypic Roles for the Endothelial Cell Receptor Kdr in Developmental Lymphangiogenesis. *Cell Rep.* **2019**, *28*, 2023–2036.e4. [CrossRef]
70. Nurmi, H.; Saharinen, P.; Zarkada, G.; Zheng, W.; Robciuc, M.R.; Alitalo, K. VEGF-C Is Required for Intestinal Lymphatic Vessel Maintenance and Lipid Absorption. *EMBO Mol. Med.* **2015**, *7*, 1418–1425. [CrossRef]
71. Fang, S.; Chen, S.; Nurmi, H.; Leppänen, V.-M.; Jeltsch, M.; Scadden, D.T.; Silberstein, L.; Mikkola, H.; Alitalo, K. VEGF-C Protects the Integrity of Bone Marrow Perivascular Niche. *Blood* **2020**, *136*, 1871–1883. [CrossRef] [PubMed]
72. Fang, S.; Nurmi, H.; Heinolainen, K.; Chen, S.; Salminen, E.; Saharinen, P.; Mikkola, H.K.A.; Alitalo, K. Critical Requirement of VEGF-C in Transition to Fetal Erythropoiesis. *Blood* **2016**, *128*, 710–720. [CrossRef] [PubMed]
73. Hasselhof, V.; Sperling, A.; Buttler, K.; Ströbel, P.; Becker, J.; Aung, T.; Felmerer, G.; Wilting, J. Morphological and Molecular Characterization of Human Dermal Lymphatic Collectors. *PLOS ONE* **2016**, *11*, e0164964. [CrossRef]
74. Facucho-Oliveira, J.; Bento, M.; Belo, J.-A. Ccbe1 Expression Marks the Cardiac and Lymphatic Progenitor Lineages during Early Stages of Mouse Development. *Int. J. Dev. Biol.* **2011**, *55*, 1007–1014. [CrossRef]
75. Schoppmann, S.F.; Birner, P.; Stöckl, J.; Kalt, R.; Ullrich, R.; Caucig, C.; Kriehuber, E.; Nagy, K.; Alitalo, K.; Kerjaschki, D. Tumor-Associated Macrophages Express Lymphatic Endothelial Growth Factors and Are Related to Peritumoral Lymphangiogenesis. *Am. J. Pathol.* **2002**, *161*, 947–956. [CrossRef]
76. Machnik, A.; Neuhofer, W.; Jantsch, J.; Dahlmann, A.; Tammela, T.; Machura, K.; Park, J.-K.; Beck, F.-X.; Müller, D.N.; Derer, W.; et al. Macrophages Regulate Salt-Dependent Volume and Blood Pressure by a Vascular Endothelial Growth Factor-C–Dependent Buffering Mechanism. *Nat. Med.* **2009**, *15*, 545–552. [CrossRef] [PubMed]
77. Harvey, N.L.; Gordon, E.J. Deciphering the Roles of Macrophages in Developmental and Inflammation Stimulated Lymphangiogenesis. *Vasc. Cell* **2012**, *4*, 15. [CrossRef] [PubMed]
78. Heino, T.I.; Kärpänen, T.; Wahlström, G.; Pulkkinen, M.; Eriksson, U.; Alitalo, K.; Roos, C. The Drosophila VEGF Receptor Homolog Is Expressed in Hemocytes. *Mech. Dev.* **2001**, *109*, 69–77. [CrossRef]
79. Seipel, K.; Eberhardt, M.; Müller, P.; Pescia, E.; Yanze, N.; Schmid, V. Homologs of Vascular Endothelial Growth Factor and Receptor, VEGF and VEGFR, in the Jellyfish *Podocoryne carnea. Dev. Dyn.* **2004**, *231*, 303–312. [CrossRef] [PubMed]
80. Tarsitano, M.; Falco, S.D.; Colonna, V.; McGhee, J.D.; Persico, M.G. The C. Elegans Pvf-1 Gene Encodes a PDGF/VEGF-like Factor Able to Bind Mammalian VEGF Receptors and to Induce Angiogenesis. *FASEB J.* **2006**, *20*, 227–233. [CrossRef]
81. Case, S.T.; Cox, C.; Bell, W.C.; Hoffman, R.T.; Martin, J.; Hamilton, R. Extraordinary Conservation of Cysteines Among Homologous Chironomus Silk Proteins Sp185 and Sp220. *J. Mol. Evol.* **1997**, *44*, 452–462. [CrossRef]
82. Johns, S.C.; Yin, X.; Jeltsch, M.; Bishop, J.R.; Schuksz, M.; Ghazal, R.E.; Wilcox-Adelman, S.A.; Alitalo, K.; Fuster, M.M. Functional Importance of a Proteoglycan Co-Receptor in Pathologic Lymphangiogenesis. *Circ. Res.* **2016**, *119*, 210–221. [CrossRef] [PubMed]
83. Keskitalo, S.; Tammela, T.; Lyytikka, J.; Karpanen, T.; Jeltsch, M.; Markkanen, J.; Yla-Herttuala, S.; Alitalo, K. Enhanced Capillary Formation Stimulated by a Chimeric Vascular Endothelial Growth Factor/Vascular Endothelial Growth Factor-C Silk Domain Fusion Protein. *Circ. Res.* **2007**, *100*, 1460–1467. [CrossRef]
84. Tammela, T.; He, Y.; Lyytikkä, J.; Jeltsch, M.; Markkanen, J.; Pajusola, K.; Ylä-Herttuala, S.; Alitalo, K. Distinct Architecture of Lymphatic Vessels Induced by Chimeric Vascular Endothelial Growth Factor-C/Vascular Endothelial Growth Factor Heparin-Binding Domain Fusion Proteins. *Circ. Res.* **2007**, *100*, 1468–1475. [CrossRef]
85. Hyytiäinen, M.; Penttinen, C.; Keski-Oja, J. Latent TGF-Beta Binding Proteins: Extracellular Matrix Association and Roles in TGF-Beta Activation. *Crit. Rev. Clin. Lab. Sci.* **2004**, *41*, 233–264. [CrossRef] [PubMed]
86. Stacker, S.A.; Stenvers, K.; Caesar, C.; Vitali, A.; Domagala, T.; Nice, E.; Roufail, S.; Simpson, R.J.; Moritz, R.; Karpanen, T.; et al. Biosynthesis of Vascular Endothelial Growth Factor-D Involves Proteolytic Processing Which Generates Non-Covalent Homodimers. *J. Biol. Chem.* **1999**, *274*, 32127–32136. [CrossRef]
87. Leppanen, V.-M.; Jeltsch, M.; Anisimov, A.; Tvorogov, D.; Aho, K.; Kalkkinen, N.; Toivanen, P.; Ylä-Herttuala, S.; Ballmer-Hofer, K.; Alitalo, K. Structural Determinants of Vascular Endothelial Growth Factor-D Receptor Binding and Specificity. *Blood* **2011**, *117*, 1507–1515. [CrossRef] [PubMed]
88. Rissanen, T.T.; Markkanen, J.E.; Gruchala, M.; Heikura, T.; Puranen, A.; Kettunen, M.I.; Kholová, I.; Kauppinen, R.A.; Achen, M.G.; Stacker, S.A.; et al. VEGF-D Is the Strongest Angiogenic and Lymphangiogenic Effector Among VEGFs Delivered Into Skeletal Muscle via Adenoviruses. *Circ. Res.* **2003**, *92*, 1098–1106. [CrossRef] [PubMed]
89. Paquet-Fifield, S.; Levy, S.M.; Sato, T.; Shayan, R.; Karnezis, T.; Davydova, N.; Nowell, C.J.; Roufail, S.; Ma, G.Z.-M.; Zhang, Y.-F.; et al. Vascular Endothelial Growth Factor-d Modulates Caliber and Function of Initial Lymphatics in the Dermis. *J. Investig. Dermatol.* **2013**, *133*, 2074–2084. [CrossRef]
90. Matsumura, M.; Bhatt, A.S.; Andress, D.; Clegg, N.; Takayama, T.K.; Craik, C.S.; Nelson, P.S. Substrates of the Prostate-Specific Serine Protease Prostase/KLK4 Defined by Positional-Scanning Peptide Libraries. *Prostate* **2005**, *62*, 1–13. [CrossRef]

91. Gupta, N.; Sudhakar, D.V.S.; Gangwar, P.K.; Sankhwar, S.N.; Gupta, N.J.; Chakraborty, B.; Thangaraj, K.; Gupta, G.; Rajender, S. Mutations in the Prostate Specific Antigen (PSA/KLK3) Correlate with Male Infertility. *Sci. Rep.* **2017**, *7*. [CrossRef]

92. Jonsson, M.; Linse, S.; Frohm, B.; Lundwall, Å.; Malm, J. Semenogelins I and II Bind Zinc and Regulate the Activity of Prostate-Specific Antigen. *Biochem. J.* **2005**, *387*, 447–453. [CrossRef] [PubMed]

93. Brown, L.F.; Yeo, K.-T.; Berse, B.; Morgentaler, A.; Dvorak, H.F.; Rosen, S. Vascular Permeability Factor (Vascular Endothelial Growth Factor) Is Strongly Expressed in the Normal Male Genital Tract and Is Present in Substantial Quantities in Semen. *J. Urol.* **1995**, *154*, 576–579. [CrossRef]

94. Obermair, A.; Obruca, A.; Pöhl, M.; Kaider, A.; Vales, A.; Leodolter, S.; Wojta, J.; Feichtinger, W. Vascular Endothelial Growth Factor and Its Receptors in Male Fertility. *Fertil. Steril.* **1999**, *72*, 269–275. [CrossRef]

95. Iyibozkurt, A.C.; Balcik, P.; Bulgurcuoglu, S.; Arslan, B.K.; Attar, R.; Attar, E. Effect of Vascular Endothelial Growth Factor on Sperm Motility and Survival. *Reprod. Biomed. Online* **2009**, *19*, 784–788. [CrossRef]

96. Saaristo, A.; Tammela, T.; Timonen, J.; Yla-Herttuala, S.; Tukiainen, E.; Asko-Seljavaara, S.; Alitalo, K. Vascular Endothelial Growth Factor-C Gene Therapy Restores Lymphatic Flow across Incision Wounds. *FASEB J.* **2004**, *18*, 1707–1709. [CrossRef]

97. Saaristo, A.; Tammela, T.; Färkkilä, A.; Kärkkäinen, M.; Suominen, E.; Yla-Herttuala, S.; Alitalo, K. Vascular Endothelial Growth Factor-C Accelerates Diabetic Wound Healing. *Am. J. Pathol.* **2006**, *169*, 1080–1087. [CrossRef] [PubMed]

98. Lawrence, M.G.; Lai, J.; Clements, J.A. Kallikreins on Steroids: Structure, Function, and Hormonal Regulation of Prostate-Specific Antigen and the Extended Kallikrein Locus. *Endocr. Rev.* **2010**, *31*, 407–446. [CrossRef]

99. Joukov, V.; Pajusola, K.; Kaipainen, A.; Chilov, D.; Lahtinen, I.; Kukk, E.; Saksela, O.; Kalkkinen, N.; Alitalo, K. A Novel Vascular Endothelial Growth Factor, VEGF-C, Is a Ligand for the Flt4 (VEGFR-3) and KDR (VEGFR-2) Receptor Tyrosine Kinases. *EMBO J.* **1996**, *15*, 290–298. [CrossRef]

100. Jeltsch, M.; Kaipainen, A.; Joukov, V.; Meng, X.; Lakso, M.; Rauvala, H.; Swartz, M.; Fukumura, D.; Jain, R.K.; Alitalo, K. Hyperplasia of Lymphatic Vessels in VEGF-C Transgenic Mice. *Science* **1997**, *276*, 1423–1425. [CrossRef] .

101. Anderson, J.; Schiffer, C.; Lee, S.-K.; Swanstrom, R. Viral Protease Inhibitors. In *Antiviral Strategies*; Kräusslich, H.-G., Bartenschlager, R., Eds.; Handbook of Experimental Pharmacology; Springer: Berlin/Heidelberg, Germany, 2009; pp. 85–110. ISBN 978-3-540-79086-0.

102. Polverino, E.; Rosales-Mayor, E.; Dale, G.E.; Dembowsky, K.; Torres, A. The Role of Neutrophil Elastase Inhibitors in Lung Diseases. *Chest* **2017**, *152*, 249–262. [CrossRef] [PubMed]

103. López-Sendón, J.; Swedberg, K.; McMurray, J.; Tamargo, J.; Maggioni, A.P.; Dargie, H.; Tendera, M.; Waagstein, F.; Kjekshus, J.; Lechat, P.; et al. Expert Consensus Document on Angiotensin Converting Enzyme Inhibitors in Cardiovascular Disease: The Task Force on ACE-Inhibitors of the European Society of Cardiology. *Eur. Heart J.* **2004**, *25*, 1454–1470. [CrossRef]

104. Wardlaw, J.M.; Murray, V.; Berge, E.; del Zoppo, G.; Sandercock, P.; Lindley, R.L.; Cohen, G. Recombinant Tissue Plasminogen Activator for Acute Ischaemic Stroke: An Updated Systematic Review and Meta-Analysis. *Lancet* **2012**, *379*, 2364–2372. [CrossRef]

105. Krebs, R.; Jeltsch, M. Die lymphangiogenen Wachstumsfaktoren VEGF-C und VEGF-D. Teil 2. Die Rolle von VEGF-C und VEGF-D bei Krankheiten des Lymphgefäßsystems. *Lymphol. Forsch. Prax.* **2013**, *17*, 96–104. [CrossRef]

106. Hartiala, P.; Suominen, S.; Suominen, E.; Kaartinen, I.; Kiiski, J.; Viitanen, T.; Alitalo, K.; Saarikko, A.M. Phase 1 Lymfactin® Study: Short-Term Safety of Combined Adenoviral VEGF-C and Lymph Node Transfer Treatment for Upper Extremity Lymphedema. *J. Plast. Reconstr. Aesthet. Surg.* **2020**, *73*, 1612–1621. [CrossRef] [PubMed]

107. Jeltsch, M.; Karpanen, T.; Strandin, T.; Aho, K.; Lankinen, H.; Alitalo, K. Vascular Endothelial Growth Factor (VEGF)/VEGF-C Mosaic Molecules Reveal Specificity Determinants and Feature Novel Receptor Binding Patterns. *J. Biol. Chem.* **2006**, *281*, 12187–12195. [CrossRef]

108. Chiu, J.; Wong, J.W.H.; Gerometta, M.; Hogg, P.J. Mechanism of Dimerization of a Recombinant Mature Vascular Endothelial Growth Factor C. *Biochemistry* **2014**, *53*, 7–9. [CrossRef]

109. Suh, S.H.; Choe, K.; Hong, S.P.; Jeong, S.; Mäkinen, T.; Kim, K.S.; Alitalo, K.; Surh, C.D.; Koh, G.Y.; Song, J.-H. Gut Microbiota Regulates Lacteal Integrity by Inducing VEGF-C in Intestinal Villus Macrophages. *EMBO Rep.* **2019**, *20*. [CrossRef]

110. Antila, S.; Karaman, S.; Nurmi, H.; Airavaara, M.; Voutilainen, M.H.; Mathivet, T.; Chilov, D.; Li, Z.; Koppinen, T.; Park, J.-H.; et al. Development and Plasticity of Meningeal Lymphatic Vessels. *J. Exp. Med.* **2017**, *214*, 3645–3667. [CrossRef] [PubMed]

111. Breslin, J.W.; Gaudreault, N.; Watson, K.D.; Reynoso, R.; Yuan, S.Y.; Wu, M.H. Vascular Endothelial Growth Factor-C Stimulates the Lymphatic Pump by a VEGF Receptor-3-Dependent Mechanism. *Am. J. Physiol.-Heart Circ. Physiol.* **2007**, *293*, H709–H718. [CrossRef] [PubMed]

112. Karkkainen, M.J.; Saaristo, A.; Jussila, L.; Karila, K.A.; Lawrence, E.C.; Pajusola, K.; Bueler, H.; Eichmann, A.; Kauppinen, R.; Kettunen, M.I.; et al. A Model for Gene Therapy of Human Hereditary Lymphedema. *Proc. Natl. Acad. Sci. USA* **2001**, *98*, 12677–12682. [CrossRef] [PubMed]

113. Martens, R.; Permanyer, M.; Werth, K.; Yu, K.; Braun, A.; Halle, O.; Halle, S.; Patzer, G.E.; Bošnjak, B.; Kiefer, F.; et al. Efficient Homing of T Cells via Afferent Lymphatics Requires Mechanical Arrest and Integrin-Supported Chemokine Guidance. *Nat. Commun.* **2020**, *11*, 1114. [CrossRef] [PubMed]

114. Jackson, D.G. Leucocyte Trafficking via the Lymphatic Vasculature— Mechanisms and Consequences. *Front. Immunol.* **2019**, *10*. [CrossRef] [PubMed]

115. Karpanen, T.; Egeblad, M.; Karkkainen, M.J.; Kubo, H.; Ylä-Herttuala, S.; Jäättelä, M.; Alitalo, K. Vascular Endothelial Growth Factor C Promotes Tumor Lymphangiogenesis and Intralymphatic Tumor Growth. *Cancer Res.* **2001**, *61*, 1786–1790.

116. Mandriota, S.J.; Jussila, L.; Jeltsch, M.; Compagni, A.; Baetens, D.; Prevo, R.; Banerji, S.; Huarte, J.; Montesano, R.; Jackson, D.G.; et al. Vascular Endothelial Growth Factor-C-Mediated Lymphangiogenesis Promotes Tumour Metastasis. *EMBO J.* **2001**, *20*, 672–682. [CrossRef] [PubMed]

117. Skobe, M.; Hawighorst, T.; Jackson, D.G.; Prevo, R.; Janes, L.; Velasco, P.; Riccardi, L.; Alitalo, K.; Claffey, K.; Detmar, M. Induction of Tumor Lymphangiogenesis by VEGF-C Promotes Breast Cancer Metastasis. *Nat. Med.* **2001**, *7*, 192–198. [CrossRef]

118. Farnsworth, R.H.; Achen, M.G.; Stacker, S.A. The Evolving Role of Lymphatics in Cancer Metastasis. *Curr. Opin. Immunol.* **2018**, *53*, 64–73. [CrossRef] [PubMed]

119. Egeblad, M.; Werb, Z. New Functions for the Matrix Metalloproteinases in Cancer Progression. *Nat. Rev. Cancer* **2002**, *2*, 161–174. [CrossRef]

120. Danø, K.; Behrendt, N.; Høyer-Hansen, G.; Johnsen, M.; Lund, L.R.; Ploug, M.; Rømer, J. Plasminogen Activation and Cancer. *Thromb. Haemost.* **2005**, *93*, 676–681. [CrossRef]

121. Harris, N.C.; Paavonen, K.; Davydova, N.; Roufail, S.; Sato, T.; Zhang, Y.-F.; Karnezis, T.; Stacker, S.A.; Achen, M.G. Proteolytic Processing of Vascular Endothelial Growth Factor-D Is Essential for Its Capacity to Promote the Growth and Spread of Cancer. *FASEB J.* **2011**, *25*, 2615–2625. [CrossRef]

122. Alitalo, A.; Detmar, M. Interaction of Tumor Cells and Lymphatic Vessels in Cancer Progression. *Oncogene* **2012**, *31*, 4499–4508. [CrossRef] [PubMed]

123. Michaelsen, S.R.; Staberg, M.; Pedersen, H.; Jensen, K.E.; Majewski, W.; Broholm, H.; Nedergaard, M.K.; Meulengracht, C.; Urup, T.; Villingshøj, M.; et al. VEGF-C Sustains VEGFR2 Activation under Bevacizumab Therapy and Promotes Glioblastoma Maintenance. *Neuro-Oncol.* **2018**, *20*, 1462–1474. [CrossRef]

124. Bergers, G.; Hanahan, D. Modes of Resistance to Anti-Angiogenic Therapy. *Nat. Rev. Cancer* **2008**, *8*, 592–603. [CrossRef]

125. Ndiaye, P.D.; Dufies, M.; Giuliano, S.; Douguet, L.; Grépin, R.; Durivault, J.; Lenormand, P.; Glisse, N.; Mintcheva, J.; Vouret-Craviari, V.; et al. VEGFC Acts as a Double-Edged Sword in Renal Cell Carcinoma Aggressiveness. *Theranostics* **2019**, *9*, 661–675. [CrossRef] [PubMed]

126. Fankhauser, M.; Broggi, M.A.S.; Potin, L.; Bordry, N.; Jeanbart, L.; Lund, A.W.; Costa, E.D.; Hauert, S.; Rincon-Restrepo, M.; Tremblay, C.; et al. Tumor Lymphangiogenesis Promotes T Cell Infiltration and Potentiates Immunotherapy in Melanoma. *Sci. Transl. Med.* **2017**, *9*, eaal4712. [CrossRef] [PubMed]

127. Song, E.; Mao, T.; Dong, H.; Boisserand, L.S.B.; Antila, S.; Bosenberg, M.; Alitalo, K.; Thomas, J.-L.; Iwasaki, A. VEGF-C-Driven Lymphatic Drainage Enables Immunosurveillance of Brain Tumours. *Nature* **2020**, *577*, 689–694. [CrossRef] [PubMed]

128. Grada, A.A.; Phillips, T.J. Lymphedema: Pathophysiology and Clinical Manifestations. *J. Am. Acad. Dermatol.* **2017**, *77*, 1009–1020. [CrossRef]

129. Mattonet, K.; Wilting, J.; Jeltsch, M. Die genetischen Ursachen des primären Lymphödems. In *Erkrankungen des Lymphgefäßsystems*; Weissleder, H., Schuchhardt, C., Eds.; Viavital Verlag: Cologne, Germany, 2015; pp. 210–229.

130. Gordon, K.; Varney, R.; Keeley, V.; Riches, K.; Jeffery, S.; Zanten, M.V.; Mortimer, P.; Ostergaard, P.; Mansour, S. Update and Audit of the St George's Classification Algorithm of Primary Lymphatic Anomalies: A Clinical and Molecular Approach to Diagnosis. *J. Med. Genet.* **2020**, *57*, 653–659. [CrossRef]

131. Newman, B.; Lose, F.; Kedda, M.-A.; Francois, M.; Ferguson, K.; Janda, M.; Yates, P.; Spurdle, A.B.; Hayes, S.C. Possible Genetic Predisposition to Lymphedema after Breast Cancer. *Lymphat. Res. Biol.* **2012**, *10*, 2–13. [CrossRef] [PubMed]

132. Finegold, D.N.; Schacht, V.; Kimak, M.A.; Lawrence, E.C.; Foeldi, E.; Karlsson, J.M.; Baty, C.J.; Ferrell, R.E. HGF and MET Mutations in Primary and Secondary Lymphedema. *Lymphat. Res. Biol.* **2008**, *6*, 65–68. [CrossRef] [PubMed]

133. Finegold, D.N.; Baty, C.J.; Knickelbein, K.Z.; Perschke, S.; Noon, S.E.; Campbell, D.; Karlsson, J.M.; Huang, D.; Kimak, M.A.; Lawrence, E.C.; et al. Connexin 47 Mutations Increase Risk for Secondary Lymphedema Following Breast Cancer Treatment. *Clin. Cancer Res.* **2012**, *18*, 2382–2390. [CrossRef] [PubMed]

134. Brouillard, P.; Boon, L.; Vikkula, M. Genetics of Lymphatic Anomalies. *J. Clin. Investig.* **2014**, *124*, 898–904. [CrossRef] [PubMed]

135. Gordon, K.; Spiden, S.L.; Connell, F.C.; Brice, G.; Cottrell, S.; Short, J.; Taylor, R.; Jeffery, S.; Mortimer, P.S.; Mansour, S.; et al. FLT4/VEGFR3 and Milroy Disease: Novel Mutations, a Review of Published Variants and Database Update. *Hum. Mutat.* **2013**, *34*, 23–31. [CrossRef]

136. Gordon, K.; Schulte, D.; Brice, G.; Simpson, M.A.; Roukens, M.G.; van Impel, A.; Connell, F.; Kalidas, K.; Jeffery, S.; Mortimer, P.S.; et al. Mutation in Vascular Endothelial Growth Factor-C, a Ligand for Vascular Endothelial Growth Factor Receptor-3, Is Associated With Autosomal Dominant Milroy-Like Primary Lymphedema. Novelty and Significance. *Circ. Res.* **2013**, *112*, 956–960. [CrossRef]

137. Balboa-Beltran, E.; Fernández-Seara, M.J.; Pérez-Muñuzuri, A.; Lago, R.; García-Magán, C.; Couce, M.L.; Sobrino, B.; Amigo, J.; Carracedo, A.; Barros, F. A Novel Stop Mutation in the Vascular Endothelial Growth Factor-C Gene (VEGFC) Results in Milroy-like Disease. *J. Med. Genet.* **2014**, *51*, 475–478. [CrossRef] [PubMed]

138. Connell, F.; Kalidas, K.; Ostergaard, P.; Brice, G.; Homfray, T.; Roberts, L.; Bunyan, D.J.; Mitton, S.; Mansour, S.; Mortimer, P.; et al. Linkage and Sequence Analysis Indicate That CCBE1 Is Mutated in Recessively Inherited Generalised Lymphatic Dysplasia. *Hum. Genet.* **2010**, *127*, 231–241. [CrossRef]

139. Betterman, K.L.; Sutton, D.L.; Secker, G.A.; Kazenwadel, J.; Oszmiana, A.; Lim, L.; Miura, N.; Sorokin, L.; Hogan, B.M.; Kahn, M.L.; et al. Atypical Cadherin FAT4 Orchestrates Lymphatic Endothelial Cell Polarity in Response to Flow. *J. Clin. Investig.* **2020**, *130*, 3315–3328. [CrossRef]

140. Katoh, Y.; Katoh, M. Comparative Integromics on VEGF Family Members. *Int. J. Oncol.* **2006**, *28*, 1585–1589. [CrossRef]
141. Rauniyar, K.; Jha, S.K.; Jeltsch, M. Biology of Vascular Endothelial Growth Factor C in the Morphogenesis of Lymphatic Vessels. *Front. Bioeng. Biotechnol.* **2018**, *6*, 7. [CrossRef]
142. El Khoury, L.; Posthumus, M.; Collins, M.; Handley, C.J.; Cook, J.; Raleigh, S.M. Polymorphic Variation within the ADAMTS2, ADAMTS14, ADAMTS5, ADAM12 and TIMP2 Genes and the Risk of Achilles Tendon Pathology: A Genetic Association Study. *J. Sci. Med. Sport* **2013**, *16*, 493–498. [CrossRef]
143. Koistinen, H.K.; Stenman, U.-H. PSA (Prostate-Specific Antigen) and other Kallikrein-related Peptidases in Prostate Cancer. In *Kallikrein-Related Peptidase. Novel Cancer Related Biomarkers*; Magdolen, V., Sommerhoff, C., Fritz, H., Schmitt, M., Eds.; deGruyter: Berlin, Germany, 2012; Volume 2, pp. 61–81. ISBN 978-3-11-030366-7.
144. Siintola, E.; Partanen, S.; Strömme, P.; Haapanen, A.; Haltia, M.; Maehlen, J.; Lehesjoki, A.-E.; Tyynelä, J. Cathepsin D Deficiency Underlies Congenital Human Neuronal Ceroid-Lipofuscinosis. *Brain* **2006**, *129*, 1438–1445. [CrossRef] [PubMed]
145. Seligsohn, U.; Lubetsky, A. Genetic Susceptibility to Venous Thrombosis. *N. Engl. J. Med.* **2001**, *344*, 1222–1231. [CrossRef] [PubMed]
146. Tsopanoglou, N.E.; Maragoudakis, M.E. Role of Thrombin in Angiogenesis and Tumor Progression. *Semin. Thromb. Hemost.* **2004**, *30*, 63–69. [CrossRef]
147. Schuster, V.; Seregard, S. Ligneous Conjunctivitis. *Surv. Ophthalmol.* **2003**, *48*, 369–388. [CrossRef]
148. Petrova, T.V.; Karpanen, T.; Norrmén, C.; Mellor, R.; Tamakoshi, T.; Finegold, D.; Ferrell, R.; Kerjaschki, D.; Mortimer, P.; Ylä-Herttuala, S.; et al. Defective Valves and Abnormal Mural Cell Recruitment Underlie Lymphatic Vascular Failure in Lymphedema Distichiasis. *Nat. Med.* **2004**, *10*, 974–981. [CrossRef] [PubMed]
149. Norrmén, C.; Ivanov, K.I.; Cheng, J.; Zangger, N.; Delorenzi, M.; Jaquet, M.; Miura, N.; Puolakkainen, P.; Horsley, V.; Hu, J.; et al. FOXC2 Controls Formation and Maturation of Lymphatic Collecting Vessels through Cooperation with NFATc1. *J. Cell Biol.* **2009**, *185*, 439–457. [CrossRef] [PubMed]
150. Ferrell, R.E.; Baty, C.J.; Kimak, M.A.; Karlsson, J.M.; Lawrence, E.C.; Franke-Snyder, M.; Meriney, S.D.; Feingold, E.; Finegold, D.N. GJC2 Missense Mutations Cause Human Lymphedema. *Am. J. Hum. Genet.* **2010**, *86*, 943–948. [CrossRef]
151. Cappello, S.; Gray, M.J.; Badouel, C.; Lange, S.; Einsiedler, M.; Srour, M.; Chitayat, D.; Hamdan, F.F.; Jenkins, Z.A.; Morgan, T.; et al. Mutations in Genes Encoding the Cadherin Receptor-Ligand Pair DCHS1 and FAT4 Disrupt Cerebral Cortical Development. *Nat. Genet.* **2013**, *45*, 1300–1308. [CrossRef]
152. Frangoul, H.; Altshuler, D.; Cappellini, M.D.; Chen, Y.-S.; Domm, J.; Eustace, B.K.; Foell, J.; de la Fuente, J.; Grupp, S.; Handgretinger, R.; et al. CRISPR-Cas9 Gene Editing for Sickle Cell Disease and β-Thalassemia. *N. Engl. J. Med.* **2020**. [CrossRef]
153. Ledford, H. CRISPR Treatment Inserted Directly into the Body for First Time. *Nature* **2020**, *579*, 185. [CrossRef]
154. Kiseleva, R.Y.; Glassman, P.M.; Greineder, C.F.; Hood, E.D.; Shuvaev, V.V.; Muzykantov, V.R. Targeting Therapeutics to Endothelium: Are We There Yet? *Drug Deliv. Transl. Res.* **2018**, *8*, 883–902. [CrossRef] [PubMed]

biology

Article

Evolutionary Analysis of Cystatins of Early-Emerging Metazoans Reveals a Novel Subtype in Parasitic Cnidarians

Pavla Bartošová-Sojková [1,*], Jiří Kyslík [1,2], Gema Alama-Bermejo [1], Ashlie Hartigan [3], Stephen D. Atkinson [4], Jerri L. Bartholomew [4], Amparo Picard-Sánchez [1,5], Oswaldo Palenzuela [5], Marc Nicolas Faber [6], Jason W. Holland [6] and Astrid S. Holzer [1]

[1] Biology Centre, Institute of Parasitology, Czech Academy of Sciences, 37005 České Budějovice, Czech Republic; jiri.kyslik@paru.cas.cz (J.K.); gema.alama@gmail.com (G.A.-B.); amparo.picard@csic.es (A.P.-S.); astrid.holzer@paru.cas.cz (A.S.H.)

[2] Faculty of Science, University of South Bohemia, 37005 České Budějovice, Czech Republic

[3] Department of Life Sciences, Natural History Museum, London SW7 5BD, UK; ash.hartigan@gmail.com

[4] Department of Microbiology, Oregon State University, Corvallis, OR 97331, USA; stephen.atkinson@oregonstate.edu (S.D.A.); jerri.bartholomew@oregonstate.edu (J.L.B.)

[5] Fish Pathology Group, Instituto de Acuicultura Torre de la Sal (IATS-CSIC), 12595 Castellón, Spain; oswaldo.palenzuela@csic.es

[6] Scottish Fish Immunology Research Centre, Institute of Biological and Environmental Sciences, University of Aberdeen, Aberdeen AB24 3UU, UK; marc.faber@moredun.ac.uk (M.N.F.); j.holland@abdn.ac.uk (J.W.H.)

* Correspondence: bartosova@paru.cas.cz; Tel.: +420-38-777-5450

check for **updates**

Citation: Bartošová-Sojková, P.; Kyslík, J.; Alama-Bermejo, G.; Hartigan, A.; Atkinson, S.D.; Bartholomew, J.L.; Picard-Sánchez, A.; Palenzuela, O.; Faber, M.N.; Holland, J.W.; et al. Evolutionary Analysis of Cystatins of Early-Emerging Metazoans Reveals a Novel Subtype in Parasitic Cnidarians. *Biology* **2021**, *10*, 110. https://doi.org/10.3390/biology10020110

Academic Editor: Hang Fai Kwok

Received: 30 December 2020
Accepted: 31 January 2021
Published: 3 February 2021

Publisher's Note: MDPI stays neutral with regard to jurisdictional claims in published maps and institutional affiliations.

Simple Summary: Cysteine protease inhibitors (cystatins) are molecules that play key protective roles in protein degradation and are involved in the immunomodulation of host responses to parasites. Little is known about the cystatin gene repertoire, evolution, and lineage-specific adaptations of early-emerging metazoans. Using bioinformatics searches, we identified orthologues of cystatins in basal animal lineages including free-living and parasite taxa. We aimed to explore whether their cystatin gene repertoire and evolution follow similar patterns recognized for derived metazoans and whether the modifications are linked to the organism's life history. We revealed that cysteine protease inhibitors from early-emerging animal groups are highly diverse, with modifications in gene organization and protein architecture. A new subtype of cystatins was discovered in the parasitic cnidarians, the Myxozoa, which has so far been only reported for a group of derived animals: trematode flukes. We set out hypotheses to describe the driving forces for the origins of this unique cystatin subtype and propose evolutionary scenarios elucidating the current existence of cystatins in the Metazoa, especially in their early-emerging lineages. Our research identified molecules for which future functional studies may help to identify their roles in host–parasite interactions and for the parasite itself.

Abstract: The evolutionary aspects of cystatins are greatly underexplored in early-emerging metazoans. Thus, we surveyed the gene organization, protein architecture, and phylogeny of cystatin homologues mined from 110 genomes and the transcriptomes of 58 basal metazoan species, encompassing free-living and parasite taxa of Porifera, Placozoa, Cnidaria (including Myxozoa), and Ctenophora. We found that the cystatin gene repertoire significantly differs among phyla, with stefins present in most of the investigated lineages but with type 2 cystatins missing in several basal metazoan groups. Similar to liver and intestinal flukes, myxozoan parasites possess atypical stefins with chimeric structure that combine motifs of classical stefins and type 2 cystatins. Other early metazoan taxa regardless of lifestyle have only the classical representation of cystatins and lack multi-domain ones. Our comprehensive phylogenetic analyses revealed that stefins and type 2 cystatins clustered into taxonomically defined clades with multiple independent paralogous groups, which probably arose due to gene duplications. The stefin clade split between the subclades of classical stefins and the atypical stefins of myxozoans and flukes. Atypical stefins represent key evolutionary innovations of the two parasite groups for which their origin might have been linked with ancestral gene chimerization, obligate parasitism, life cycle complexity, genome reduction, and host immunity.

Keywords: cysteine protease inhibitor; stefin; signal peptide; parasite; phylogenetic analysis; diversification; protein structure

1. Introduction

Members of the cystatin superfamily (I25), also known as cystatins, are major regulators of the activity of papain-like cysteine proteases and legumain-type proteases. These endogenous protease inhibitors are widely distributed in all domains of life. Based on sequence homology, the cystatin superfamily comprises three main sub-families: types 1–3. Type 1 cystatins (stefins, I25A) are low-molecular-weight intracellular single-domain proteins generally lacking disulfide bonds and signal peptides. Within the stefins, an additional subtype, unusually featuring a signal peptide, is recognized exclusively in liver and intestinal flukes (Trematoda: Digenea) [1–4]. Type 2 cystatins (I25B) are mainly extracellular single-domain proteins containing two disulfide bonds and a signal peptide, while type 3 cystatins (kininogens, I25C) are large multi-domain proteins [5–7].

Cystatins are involved in the regulation of key physiological processes, such as the protection of cells from proteolysis [5–7]. In parasites, cystatins are involved in immunomodulation of host cells and represent important immunogenicity and pathogenicity factors involved in host–parasite interactions [8–13]. They facilitate parasite survival within the host [14,15] by inhibiting host proteases and by interfering with antigen processing and presentation [16–19]. The increased capacity of parasite cystatins to induce anti-inflammatory cytokine IL-10 [16,17,20] was suggested to be an adaptation to the parasitic lifestyle [21,22]. These molecules have been used as effective markers in diagnosis and vaccine development and as chemotherapeutic targets against various parasitic infections [23–27]. Recently, cystatins have also been suggested as candidates for immunotherapies for acute and chronic inflammatory diseases [28] and cancer [29].

The diversity and evolution of the cystatin superfamily is well-documented in prokaryotes and certain eukaryotic lineages [30–33]. Within the Metazoa, phylogenetic affinities and representation of cystatins have been investigated predominantly in the derived lineages (Bilateria) including both free-living [34] and parasite taxa, e.g., various helminth [2,10,35–38] and arthropod groups [39–42]. According to the current knowledge of cystatin evolution, the stefin and type 2 cystatin lineages were already present in the eukaryotic ancestor. Stefins have remained structurally similar to the ancestral form and are present across most metazoan groups. In contrast, type 2 cystatins greatly diversified, primarily in derived metazoans (Vertebrata), and have been lost entirely in some early (Placozoa) and derived (Acoela, Tardigrada, Hemichordata, etc.) metazoan groups. Multi-domain cystatins have been reported exclusively from several bilaterian groups [30]. Nevertheless, the picture of cystatin evolution in the Metazoa is incomplete, as early-emerging animal lineages (Porifera, Ctenophora, Placozoa, and Cnidaria) are greatly underexplored in this respect and completely underrepresented with regard to parasitic lineages. To date, only a few early-emerging animal taxa were included in the studies of cystatin evolution [30,34], from which none were parasites. Thus, further data are needed from these ancient animal lineages, including those with parasitic lifestyles, to improve our evolutionary understanding and to explore potential linkage to the organism's life history.

Basal metazoans mostly comprise free-living species, but at least eight independent parasite lineages of early-emerging Metazoa are known, with all but one in Cnidaria [43]. Recently, next-generation sequencing (NGS) data have become available for three of these parasite lineages, i.e., two *Edwardsiella* spp. [44,45], *Polypodium hydriforme* [46,47], and several species of the large group Myxozoa [47–54]. While *Edwardsiella* and *Polypodium* parasitize their hosts exclusively during the larval stage of their development [45,55], the microscopic myxozoans are an entirely parasitic group with complex life cycles [56]. Members of the specious class Myxosporea are primarily known from fish (intermediate hosts) and annelids (definitive hosts). The more primitive Malacosporea use fish and freshwater bryozoans as hosts [57]. Myxozoans are an evolutionary old and fast-evolving group, evidenced by their highly derived morphology and seen in genome phylogenies [46,47,51,53,58]. Their genomes are the smallest amongst all animals due to evolutionary transition to parasitism and the associated extreme reduction of body complexity from free-living cnidarians [47].

To explore the gene repertoire, gene organization, and structural features of cystatins in phyla at the base of the metazoan tree, we mined 110 genomes and transcriptomes of 58 early-emerging animal species. We conducted phylogenetic analyses using newly obtained and published cystatin sequences of basal and derived metazoans to provide insights into the origin, evolution, and structural diversification of this group of inhibitors at the time when this superfamily emerged and expanded. We investigated the amino acid sequence structure and phylogenies of cystatins from free-living and parasite taxa to identify structural modifications, which potentially represent lineage-specific adaptations related to parasitic lifestyles. Our investigation thus aimed to reveal whether gene repertoire and evolution of cystatins in early-emerging animals follow similar patterns recognized for derived metazoans and whether the modifications are linked to the organism's life history.

2. Materials and Methods

2.1. Data Mining

Representatives of basal metazoan lineages (Placozoa, Porifera, Ctenophora, and Cnidaria) were selected for mining of cystatin homologues from the NGS data available from GenBank, other public databases, or produced in our laboratories (110 datasets of 58 species; Table S1). The samples included free-living and parasite taxa, particularly Cnidaria, i.e., *Polypodium hydriforme*, *Edwardsiella lineata*, *E. carnea*, and all available myxozoans (13 species). We used published the cystatin homologues of metazoans from basal animal species [30,53] for the cystatin homology search. If no homologues were found in a species in the first blast, a repeated search was applied by utilizing the sequences of homologues found in related species in the first iteration as additional queries for the next search.

We used a combined search strategy using both motif- and sequence similarity-based methods to identify target molecules. The search was performed using the tBLASTn algorithm with the E-value cutoff set to 10^{-5}. This relatively low threshold was selected based on low sequence similarity among our queries and subjects (approximately 30–40%). The mined sequences were examined for the presence of typical conserved motifs of the cystatin domain (G; Q-x-V-x-G; LP/PW; YF [30,59]), by searches against different databases (GenBank, MEROPS peptidase database [60]) including the reciprocal blast analysis, and by their positioning in our preliminary phylogenetic analyses. Detailed homology searches were performed using the hmmrscan algorithm of HmmerWeb version 2.32.0 software [61] against protein family databases: Pfam, CATH-Gene3D, PIRSF, Superfamily, TIGRFAM, and TreeFam. For parasites, the potential contaminating sequences of host or other origin were identified and filtered out by blast matches with reference NGS data and by phylogenetic clustering with contaminant species homologues.

2.2. Verification of Selected Cystatin Sequences

Full gene sequences of the cystatin superfamily homologues mined from unpublished genomes and transcriptomes (Table S1) were verified by PCR and by sequencing from DNA and/or cDNA templates using species-specific primers designed in this study, if possible, in the 5′ and 3′ UTR regions (details in Methods S1 and Table S2). PCR was used to obtain the stefin gene sequence of *Buddenbrockia plumatellae* (Methods S1 and Table S2) as mining of the published *Buddenbrockia* EST database (Table S1) did not return any cystatin hits. PCR-verified sequences were deposited in GenBank under the accession Nos. MT127416–MT127426 and MW498387–MW498390 (Table S2). The positions of the introns were assessed by comparison of the sequences of DNA vs. cDNA origin and, if possible, by comparing sequences mined from the genomic vs. transcriptomic data.

2.3. Phylogenetic Analyses

We compiled and analyzed a sequence dataset (128 taxa) of both newly identified and published cystatins. The dataset comprised Choanoflagellates (11 taxa), early-emerging animal lineages (58 taxa), more derived metazoans (58 taxa), and *Giardia* cystatin (outgroup) resembling the most ancestral eukaryotic cystatin [30] (Table S3). Ingroup taxa were represented by free-living species (72 taxa) and parasite species from basal metazoans (*Polypodium hydriforme*, *Edwardsiella* spp., and myxozoans; 16 taxa) and bilaterians (trematodes, cestodes, nematodes, monogeneans, ticks, and mites; 39 taxa) (Table S3). Multi-domain cystatins were omitted from the phylogenetic analyses due to difficulties with their alignment.

We aligned 329 amino acid cystatin sequences in MAFFT v7.017 implemented in Geneious v8.1.3 [62] using the iterative refinement by the G-INS-i strategy that resulted in an alignment of stefins and type 2 cystatins according to topological equivalence in tertiary structure (see Dataset S1 of [59]). Nonhomologous regions at the beginning (e.g., signal peptides) and end of the alignment were trimmed so the final alignment comprised 213 positions, principally corresponding to the conserved cystatin domain (Pfam: PF00031) (Dataset S2). The phylogenetic trees were reconstructed using the maximum likelihood (ML) and Bayesian inference (BI) methods. ML analysis was performed in RAxML v7.0.3 [63], using the WAG + G model selected by ModelFinder implemented in IQ-TREE webserver [64]. Indels were treated as unknown characters. Bootstraps were based on 1000 replicates. BI analysis was performed in MrBayes v3.2.7av1 [65] using the WAG model of evolution, with 24 million generations sampled at intervals of 1000 trees until the standard deviation from split frequencies was below 0.05. The burn-in period represented the default initial 25% of all generations. The trees were visualized in Geneious v8.1.3 and graphically modified in Adobe Illustrator CS5.

2.4. Comparison of the Protein Architecture

Cystatin diversity on the amino acid sequence level was analyzed in terms of critical motifs, amino acid conservation, presence/absence of a signal peptide and disulfide bridges, and modifications that might potentially affect the cystatin inhibitory activity. Disulfide bonds were predicted using the DISULFIND webserver [66]. Signal peptides were predicted using the SignalP v5.0 webserver [67]. Animal group-/cystatin type-specific sequence motifs were identified by comparison of primary structures of cystatin amino acid sequences in the WebLogo v3 software [68] using extractions of the alignment used for the phylogenetic analyses.

3. Results

In this study, we retrieved 184 homologues of cystatins (104 stefins and 80 type 2 cystatins) from the NGS data of early-emerging metazoans (Table S1). Of these, we verified with PCR the identity of 15 homologues mined from unpublished datasets (Table S2).

3.1. Cystatin Gene Repertoire and Diversity in Early-Emerging Metazoans

We observed a large variation in the representation of cystatin subclasses in the investigated animal groups (Figure 1). Type 2 cystatins were missing in several groups (Placozoa, Porifera, Cnidaria: Staurozoa, and Myxozoa). Stefins were present in all but one (Placozoa) basal metazoan lineage. We found a separate group of stefins in myxozoan parasites, designated here as "atypical stefins", which differed in sequence structure to classical metazoan stefins (see Section 3.3 and Figure 2). No multi-domain cystatins could be identified in basal metazoans (Figure 1).

Taxonomic group	Classical stefins	Atypical stefins	Type 2 cystatins	Multi-domain cystatins
Placozoa	O	O	O	O
Porifera	●	O	O	O
Ctenophora	●	O	●	O
Cnidaria/Anthozoa (+ *Edwardsiella* spp.*)	●	O	●	O
Cnidaria/Staurozoa	●	O	O	O
Cnidaria/Cubozoa	●	O	●	O
Cnidaria/Scyphozoa	●	O	●	O
Cnidaria/Hydrozoa	●	O	●	O
Cnidaria/Polypodiozoa*	●	O	●	O
Cnidaria/Myxozoa**	O	●	O	O

Taxa parasitic only in larval stage of their life cycle (*) or throughout the whole life cycle (**)
Feature absent (**O**) or present (**●**).

Figure 1. Repertoire of cystatin superfamily genes in early-emerging metazoan lineages.

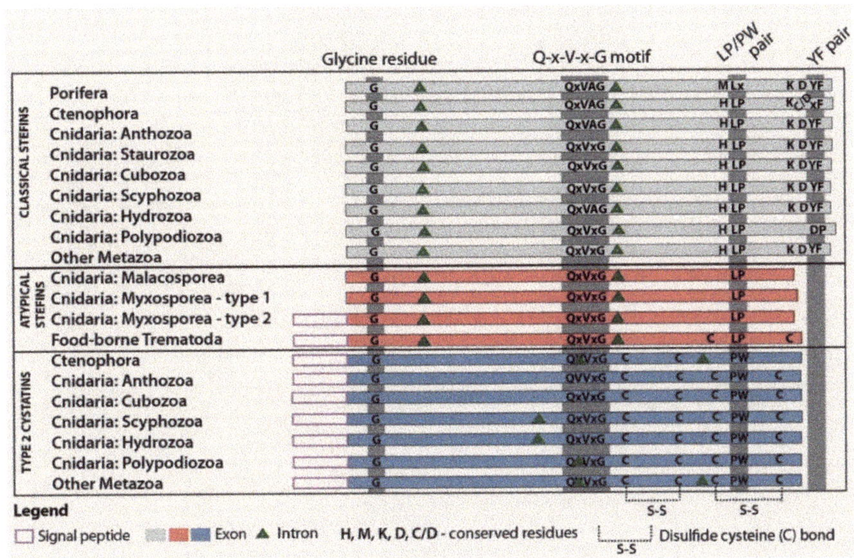

Figure 2. Schematic comparison of the gene architecture and amino acid structure of metazoan cystatins: for myxozoans, information is shown separately for Malacosporea and Myxosporea.

We found that basal metazoans had a diverse cystatin gene repertoire: most investigated species had multiple cystatin homologues with significant sequence differences, even on the intraspecies level (i.e., several divergent sequences of type 2 cystatin and/or stefin homologues in the genome of a single species; Table S1). Exceptionally, certain species (e.g., cnidarians *Ceratonova shasta* and *Physalia physalis*) only possessed single type 2 cystatin and/or stefin homologue.

All mined cystatin homologues are listed either as new GenBank entries (for PCR-verified sequences mined from data unpublished at the time of the homology searches) or as Genbank contig/scaffold identifiers for sequences mined from published data (Table S2 and Dataset S1).

3.2. Phylogeny of Metazoan Cystatins did not Mirror Animal Phylogeny and Was Highly Diverse within Individual Groups

The clustering of cystatins in both ML and BI analyses did not follow the generally accepted trends of organismal phylogeny, as early-emerging animal lineages did not occupy basal positions in our trees (Figure 3 and Figure S1) as commonly known from other phylogenetically informative genes., e.g., 18S rDNA. Based on differences in the primary amino acid sequence structure, cystatins clustered into a single weakly supported stefin clade (ML/BI = 48/0.75) while the remainder of sequences clustered within the type 2 cystatin clade (ML/BI = 59/0.92). The stefin clade included a mixture of classical and atypical stefin subclades with weakly supported interrelationships. Classical stefin subclades comprised all Metazoa except the myxozoans, while atypical stefin subclades were formed exclusively of certain early-emerging (myxozoans) and more derived (trematodes) parasite animal lineages. Unlike other metazoan stefins, those of myxozoans were extremely divergent in their sequences, thus creating long branches (Figure 3 and Figure S1).

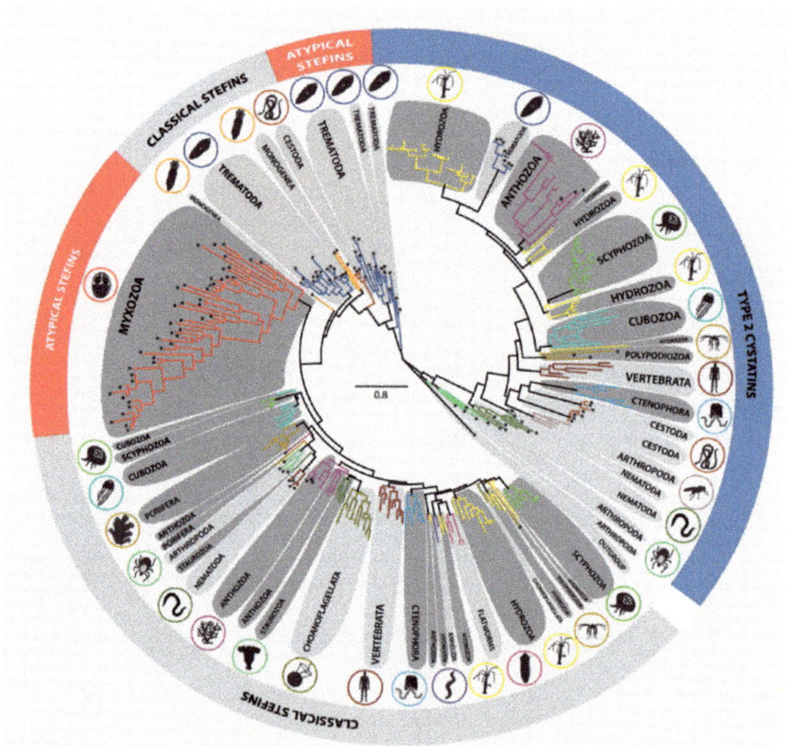

Figure 3. Maximum likelihood phylogenetic tree of the cystatin superfamily represented in the Metazoa and Choanoflagellata. *Giardia* cystatin was used as the outgroup. The basal metazoan groups are depicted in dark grey shaded bubbles, while other groups are shaded light grey. Cystatin homologues from parasite taxa are labelled by asterisks at the tip of the nodes. The detailed ML and BI phylogenetic trees featuring taxa names and all nodal supports are provided in Figure S1.

Both stefins and type 2 cystatins clustered into subclades that paralleled classic taxonomy. However, interclade relationships of these taxonomy-defined subclades were weakly supported/not resolved due to polytomies (Figure 3 and Figure S1) and did not always reflect the organismal phylogeny. Indeed, some representatives formed monophyletic taxonomic groups, e.g., Ctenophora, Porifera, Cubozoa, and Myxozoa, while some cnidarian groups (Polypodiozoa, Staurozoa, Scyphozoa, Anthozoa, and Hydrozoa) split into multiple subclades not related by classic taxonomy. We attributed this splitting to (i) sequence differences among species of the same taxonomic group (e.g., hydrozoans *Hydra*, *Podocoryna*, *Physalia*, *Velella*, and *Porpita* spp. clustering into independent type 2 cystatin clades) or (ii) the occurrence of paralogues (e.g., stefin out-paralogues of the scyphozoan *Aurelia aurita* present in two distinct groups). Some of the out-paralogous clades additionally contained multiple sequences from single species (e.g., stefin in-paralogues of *A. aurita* ranging up to 37.7% amino acid sequence divergence) (Figure 3 and Figure S1).

3.3. Structural Modifications of Cystatin Homologues in Basal Metazoans

The results of structural analyses aimed at identifying intron number and positions, conserved regions, and the presence/absence of signal peptides and cysteine bonds were combined to create the comparative scheme of cystatin structural motifs (Figure 2).

We identified that introns in cystatin nucleotide sequences of early-emerging animal lineages varied in number (0–2) and location in type 2 cystatins while stefins contained two introns with conserved positions (Figure 2). All introns were delimited by typical GT/AG sequence boundaries at the exon–intron junction.

In amino acid sequence data from our alignment of basal animals (Dataset S2), we observed four conserved regions important for interaction with proteases, as found previously with derived metazoans [30,59]: (i) a glycine residue within the amino-terminal "trunk", (ii) a Q-x-V-x-G motif within the first hairpin loop, (iii) the PW (type 2 cystatins) or LP (stefins) motifs within the second hairpin loop, and (iv) a tyrosine residue located in the relatively conserved D-x-L-x-Y-F carboxy terminus of stefins. We observed two novel conserved residues in stefins of both early-emerging and derived metazoans: (i) a histidine residue located seven amino acids before the LP pair towards the N-terminus and (ii) a lysine residue situated eight amino acids before the C-terminal conserved tyrosine residue (Figure 2 and Dataset S2).

While the amino acid structure of the first two key activity regions (G, Q-x-V-x-G) was conserved among all sequences, several modifications were seen in other regions. All sponge (Porifera) stefins showed a characteristic amino acid replacement of the LP pair to LD (or LS in one case). Substitution of the conserved C-terminal tyrosine residue by non-aromatic valine was observed in two homologues. The conserved histidine residue located near the LP pair was commonly replaced by methionine or, exceptionally, by histidine or tryptophan (both observed once) (Dataset S2).

Stefins and type 2 cystatins of comb jellies (Ctenophora) had the typical sequence motifs defined for these molecules in other metazoans. In a few taxa, we observed substitutions of the conserved cystatin PW pair by AW and stefin C-terminal tyrosine residue by proline or histidine (Dataset S2).

Cnidarian cysteine protease inhibitors showed a wide variety of modifications (Figure 2 and Dataset S2). Cubozoans were the only group for which conserved stefin regions had the classical organization, and their type 2 cystatins differed in a few cases by replacement of the PW motif with SW or KF. Stefins of hydrozoans and scyphozoans showed single substitutions of LP to FA or LK and single alterations of C-terminal tyrosine to aromatic phenylalanine, and many of the type 2 cystatins had substitutions of PW with AW, RW, or SW and exceptionally with PF or PL. Staurozoan stefins had C-terminus modifications, where the tyrosine residue was substituted with histidine or was missing. In anthozoans, type 2 cystatins had alterations of PW to SW or GW, while a few stefins had substitutions of LP with FD or FS and replacement of the C-terminal tyrosine residue to proline or histidine.

The most notable modification we found was in two parasitic cnidarian groups Polypodiozoa and Myxozoa. While the majority of type 2 cystatins of *P. hydriforme* had the typical conserved motifs, its stefin was highly modified, with the LP pair substituted with LS and the C-terminal conserved D-x-L-x-Y-F motif replaced with a completely different pattern lacking a tyrosine residue. While this stefin retained the histidine found near the LP pair, the conserved lysine located near the C-terminal was replaced by leucine. Myxozoans, as representatives of early-emerging metazoans, had atypical stefins, which completely lacked the D-x-L-x-Y-F region at the C-terminus and had the normally conserved histidine and lysine residues replaced by other residues. Stefins of an evolutionary older myxozoan subgroup, the Malacosporea, retained the original LP pair, but some stefins of the specious and more derived myxozoan group, the Myxosporea, had replaced this motif with LS, LY, LR, FK, or SA (Figure 2). We could not find classical stefins or type 2 cystatins in any myxozoan (Figure 1).

In general, signal peptides in cystatins of basal metazoan lineages followed established patterns: they were present in type 2 cystatins and absent in classical stefins. The only exceptions were some myxosporean stefins that carried a signal peptide (Figure 2). Interestingly, the pattern of signal peptide absence (in type 1 atypical stefins) or presence (in type 2 atypical stefins) was observed even at the intraspecies level. For example, some myxosporeans coded exclusively for stefins without a signal peptide (i.e., *C. shasta* and *Sphaeromyxa zaharoni*) while others presented multiple stefin genes that all carried the signal peptide (i.e., *Myxobolus pendula* and *M. cerebralis*) or had genes that represented both variations (i.e., *Enteromyxum leei, E. scophthalmi, Kudoa iwatai, Thelohanellus kitauei, Myxidium lieberkuehni* and *Sphaerospora molnari*) (Figure 2 and Dataset S1).

As typical for type 2 cystatins, two conserved disulfide bridges were identified in all basal metazoans with the exception of homologues found in ctenophorans, which were predicted to contain only one disulfide bridge. Classical stefins lacked disulfide bridges, as is typical for this cystatin subfamily type, and this pattern was also observed in atypical stefins (Figure 2).

3.4. Structural and Evolutionary Comparison of Cystatins of Parasitic vs. Free-Living Groups of Basal Metazoans: The Special Case of Myxozoa

The Myxozoa, the only obligate parasite group of early-emerging metazoans, possessed exclusively atypical stefins and lacked type 2 cystatins. Myxozoan stefins had characteristic differences in sequence to those of the other basal Metazoa (see Section 3.3) and clustered into a single clade with long branches (see Section 3.2). The two cnidarian lineages studied that are parasitic only in their larval stage (*E. lineata, E. carnea,* and *P. hydriforme*) possessed both classical stefins and type 2 cystatins, and no atypical stefins were found in these animals. Their stefins formed a distinct lineage that afterwards clustered with other cnidarians (*P. hydriforme*) or grouped with their respective anthozoan members (*E. lineata* and *E. carnea*). Similarly, clustering of "partially parasitic" cnidarians into distinct monotypic lineages (*P. hydriforme*) or within anthozoan lineages (*E. lineata* and *E. carnea*) was observed for type 2 cystatins (Figure 3 and Figure S1). No structurally unique molecules were identified in the entirely free-living basal metazoans.

4. Discussion

4.1. A Diverse Repertoire of Cystatins in Early-Emerging Metazoans

In this study, we identified the cystatin gene repertoire of early-emerging animals by mining homologues from 110 genomes and transcriptomes of basal Metazoa. We demonstrate that cystatins are widely distributed and highly diversified in current representatives of these ancient lineages and that the repertoire of cystatin superfamily genes differs widely among phyla. Each taxonomic group evolved multiple stefin and/or type 2 cystatin homologues, forming individual taxonomy-defined clades in the phylogeny. These paralogues probably arose by gene duplication [30] rather than by whole genome multiplication, as the only basal metazoans for which genome duplication has been suggested are members of the cnidarian genus *Acropora* [69]. Multiple out-paralogues found in cnidarians likely originated before cnidarian radiation while, in-paralogues, probably emerged after the speciation event. Bursts of sequence diversification likely gave rise to gene orthologues within each taxonomy-defined clade. These evolutionary processes resulted in the origin of diverse molecules with varying inhibitory and immunomodulatory functions in metazoans.

Some taxonomic groups had a wide diversity of cystatins, while others lacked some or all cystatin subfamilies. For example, genes coding for cystatins were not identified in the placozoan genomes: two different strains of *Trichoplax* sp. [70] and *Hoilungia hongkongensis* [71], which was in concordance with findings of Kordiš and Turk [30] from the related species *Trichoplax adhaerens* [70,72]. Evidently, cystatins are not essential for the survival of these simple organisms despite the presence of cysteine proteases in them [71,72]. We suggest that, in placozoans, the functions of cystatins have been replaced by other types of inhibitors, such as structurally unrelated equistatin or serpins [70–72]. These proteins have been recognized in other organisms as potent inhibitors of cysteine proteases [73] or cross-class inhibitors of them [74].

4.2. Atypical Stefins: A Unique Type of Cystatins in Myxozoans and Some Trematodes

Interestingly, we found a wide diversity of both classical and atypical stefins in basal metazoans and we suspect that these highly divergent homologues may have different or novel functions, as suggested for the highly divergent myxozoans serpins (serine protease inhibitors [75]). Stefins structurally similar to those of myxozoans have been identified only from more derived bilaterians: intestinal and liver flukes (Trematoda: Digenea) [1–4]. While myxozoans possess only atypical stefins, these trematodes have cystatins from several superfamily subclasses (classical and atypical stefins, multi-domain cystatins [1–4,22,76]). The atypical stefins presumably arose independently in the two groups, possibly as a result of pressures from a parasitic life history; however, other parasites do not possess these molecules. For example, parasitic metazoans (monogeneans, nematodes, cestodes, ticks, and mites) [9,36,37,39,42] including blood trematodes [35] have only classical stefins and type 2 cystatins. Similarly, cnidarians *Polypodium* and *Edwardsiella* spp. with parasitic larval and free-living adult stages also have only classical stefins and type 2 cystatins. Thus, atypical stefins represent key evolutionary innovations of myxozoans and intestinal and liver trematodes. The likely function of atypical stefins in myxozoans and trematodes is to control unwanted proteolysis by cysteine proteases expressed by different parasite life stages and by the host [1].

Structurally, atypical stefins contain the characteristic stefin LP pair, which is modified in certain myxozoan stefins. Importantly, all atypical stefins lack the conserved D-x-L-x-Y-F carboxy-terminal part that is generally found in classical stefins but is absent from type 2 cystatins (Figure 2). As the tyrosine residue of the D-x-L-x-Y-F region is important for stabilization of the stefin-protease complex [59,77], the absence of this motif along with the modifications in the conserved LP pair may have functional implications in atypical stefins, e.g., their capability to inhibit a broader range of proteases, including those of host origin. The function of this C-terminal modification warrants future experimental studies as demonstrated previously for stefin mutants [77,78].

Additionally, we identified that the atypical stefins of some myxozoans and all trematodes have a signal peptide typical for type 2 cystatins (Figure 2). In some taxa, this feature correlates with extracorporeal protein secretion [3,79–81], but this is not consistent across the Metazoa, as not all proteins that have a signal peptide are exported out of the organism [1] and a large proportion of excreted/secreted products do not have a signal peptide [1,4,37,79,81–83]. Secreted proteins, including cystatins, are crucial for parasite protection from degradation by host cysteine proteases and for modulation of host immunity [9–11,16,79]. We postulate that some myxozoan stefins are secretory proteins with immunomodulatory roles, as has been suggested for the atypical stefin of the myxosporean *Thelohanellus kitauei* [53,84]. These molecules are likely responsible for the increased expression of the anti-inflammatory IL-10 gene in fish hosts infected with different myxosporean species [85–90] as an increased production of IL-10 in hosts infected by other parasites is elicited by cystatin-induced immunomodulation [16,20]. Similar to other parasites [23–27], we suggest that these proteins should be further explored in myxozoans as vaccine targets due to the lack of efficient strategies against these parasites in fish destined for human consumption [91,92].

We propose two mechanisms for the origin of atypical stefins: (i) structural modification of ancestral classical stefins by the acquisition of a signal peptide in some homologues and loss of the C-terminal and its tyrosine residue, or (ii) combination of the features of classical stefins (LP pair) and type 2 cystatins (signal peptides, missing C-terminal part) in chimeric genes. Chimeric genes are important in the evolution of genetic novelty and can allow organisms to adapt to novel environments (or hosts) by acquisition of novel functions [93,94]. Both myxozoans and trematodes are obligate parasites with complex life cycles that involve vertebrate and invertebrate hosts [8,95]. The independent origins of their atypical stefins might have been associated with the transition of a presumably free-living ancestor to obligate parasitism and the associated challenges of surviving within one or more hosts. Extending a recent hypothesis that invertebrates were the initial hosts for myxozoans and fish were incorporated into life cycles later [96], we postulate that myxozoans reduced their type 2 cystatins in the invertebrate host in the course of genome reduction due to parasitism [47] since invertebrates have limited immune capacity [97]. Ancestral superfamily cystatin types were lost in myxozoans as their roles were replaced by atypical stefins, which in trematodes have been shown to inhibit both parasite and host proteases [2,3]. Then, under novel pressures in adopted fish hosts, myxozoans remodeled classical stefins into the atypical ones to interfere with specific immune responses of fish and to enable novel behaviors such as migration within host tissues.

4.3. Obstacles to Reconstruction of Cystatin Phylogenies

In our analyses, the reconstruction of evolutionary relationships in the cystatin superfamily had good support for the more recent clades but not for deeper evolutionary nodes, as reported previously for this group of genes [30,35]. Unstable topology, polytomies (in BI), and low nodal supports in our phylogeny were due to low informative content from short protein length and high sequence divergence in the chosen genes [30,35]. However, despite generally poor support, a clear separation of ingroup taxa into distinct lineages defined by taxonomy and protein architecture revealed remarkable trends in the evolution of cystatins in early-emerging metazoans, with structurally unique molecules of potentially novel functions. Different branches appear to have evolved at different rates [98], especially in fast-evolving myxozoans [46,47,51,53,58]. Fast rates of evolution of the myxozoan genes likely caused the extraordinary sequence divergence of myxozoan stefins, which may represent functional diversification of these molecules. Phylogenetic clustering of atypical stefins within the myxozoan clade seems to some extent to follow myxozoan speciation and differentiation into the well-known myxozoan lineages (oligochaete-infecting, polychaete-infecting, *Sphaerospora sensu stricto*, and malacosporeans [96]) Resolution of these established myxozoan clades is low due to the undersampling of taxa and is probably compounded by diversification including a wide functional range of atypical stefins, with or without signal peptides, and with each particular lineage displaying specific sequence motifs and potential host organ-specific protease expression.

4.4. Hypothetical Evolution of Metazoan Cystatins

From their comprehensive analysis of prokaryotic and eukaryotic cystatins, Kordiš and Turk [30] concluded that the ancestor of the cystatin superfamily was intracellular and lacked a signal peptide and disulfide bridges (similarly to extant stefins). Initial gene duplication produced two ancestral eukaryotic lineages, stefins and type 2 cystatins, with the latter evolving from the stefin-like ancestor by acquiring cysteine residues, disulfide bridges, and a signal peptide. Later gene and domain duplications combined with deletions and insertions of genetic material resulted in diverse single- and multi-domain proteins with or without disulfide bonds and glycosylations [30].

Based on our novel data and published records [30], we propose the following alternative scenarios for the evolution of single-domain cystatins in the Metazoa, with a focus on its basal lineages (Figure 4). After the split of the cystatin superfamily into the initial stefin and type 2 cystatin lineages in the ancestor of eukaryotes, stefins were retained in the common ancestor of the Metazoa and all its descendant lineages except Placozoa. Sequence diversification and lineage-specific adaptation of stefins occurred independently in parasites (Myxozoa and Trematoda), giving rise to atypical stefins. Type 2 cystatins were either (i) retained in the metazoan ancestor and its descendant lineages except for their independent loss in Porifera, Placozoa, some Cnidaria (Myxozoa, Staurozoa) and some bilaterians (liver and intestinal flukes) (scenario A in Figure 4); (ii) initially lost in the metazoan ancestor (or even before), reappeared in the ancestor of Eumetazoa, and were again lost in some descendant lineages (Placozoa, Myxozoa, Staurozoa, and some flukes) (scenario B in Figure 4); or (iii) initially lost in the metazoan ancestor (or even before) and reappeared independently in Ctenophora and in the ancestor of the lineage, leading to Cnidaria and Bilateria with some groups (Myxozoa, some flukes) having a secondary loss (scenario C in Figure 4).

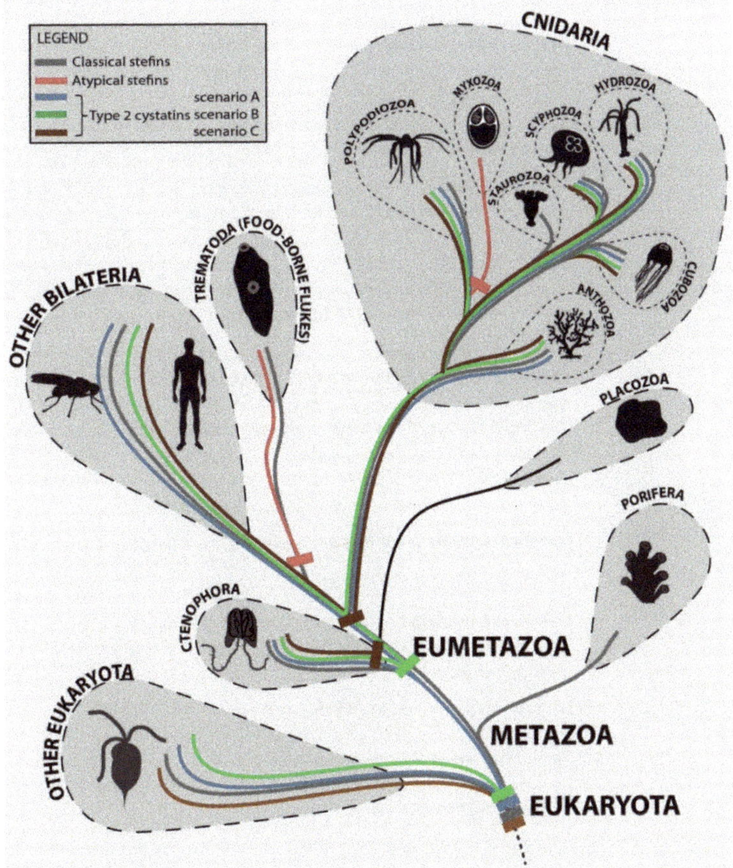

Figure 4. Hypothetical evolutionary scenarios of origins of stefins and type 2 cystatins in the Metazoa with a focus on early-emerging animal lineages: please note that the different colors for type 2 cystatins visualize the alternative scenarios of hypothetical evolution for the same cystatin subtype and not the sequence differences.

5. Conclusions

Stefins are considered relatively conserved genes with characteristics closer to the ancestral superfamily type, while type 2 cystatins are considered to have undergone a more complex and dynamic evolution through numerous gene and domain duplications [30]. In this study, we showed that early-emerging metazoan lineages were more prone to type 2 cystatin loss and that significant diversification of this type of molecules occurred only later in their evolution. Simultaneously, these ancient lineages retained stefin genes that were more dynamic with regard to structural modifications, with bursts of diversification creating atypical stefins in some obligate parasites with complex life cycles. Further research aimed at biochemical and functional characterization of the parasite cysteine protease inhibitors, especially of atypical stefins, is necessary to decipher the roles of these proteins in multi-host–parasite interactions. This knowledge will improve the understanding of the contributions of cystatins for parasite survival in the host causing disease and may reveal targets for therapeutant development against Myxozoa.

Supplementary Materials: The following are available online at https://www.mdpi.com/2079-7737/10/2/110/s1, Figure S1: Maximum likelihood and Bayesian trees with nodal supports; Table S1: Next-generation sequencing databases mined for cystatins; Table S2: List of myxozoan species with information on primers used for the PCR verification of stefin genes; Table S3: The list of groups/species used for phylogenetic analyses. Methods S1: Methodological details regarding the PCR verification of myxozoan stefins; Dataset S1: Fasta file of complete the amino acid sequences of cystatins present in the phylogenetic analyses; Dataset S2: Fasta-formatted alignment of the trimmed amino acid sequences of cystatins used for the phylogenetic analyses.

Author Contributions: Conceptualization, P.B.-S.; methodology, P.B.-S.; formal analysis: P.B.-S., J.K., O.P., M.N.F., and G.A.-B.; investigation: P.B.-S., S.D.A., O.P., A.P.-S., A.H., M.N.F., and G.A.-B.; resources: P.B.-S., A.S.H., O.P., J.W.H., and J.L.B.; data curation: P.B.-S. and O.P.; writing—original draft preparation: P.B.-S.; writing-review and editing: All authors; visualization: P.B.-S. and J.K.; supervision: P.B.-S.; funding acquisition: P.B.-S. and A.S.H. All authors have read and agreed to the published version of the manuscript.

Funding: This research was funded by the Ministry of Education, Youth, and Sports of the Czech Republic, grant number LTAUSA17201; by the European Commission under the H2020 Programme—ParaFishControl, grant number 634429; by the Czech Science Foundation, grant number 19-28399X (to A. S. Holzer, G. Alama-Bermejo, and J. Kyslík) and 21-16565S and by the Czech Academy of Sciences and Hungarian Academy of Sciences, grant number MTA 19-07. This publication reflects the views of the authors only; the European Commission cannot be held responsible for any use which may be made of the information contained therein.

Institutional Review Board Statement: Not applicable.

Informed Consent Statement: Not applicable.

Data Availability Statement: The data presented in this study are openly available in Genbank (under the accession Nos. MT127416–MT127426 and MW498387–MW498390) and in the associated supplementary files.

Acknowledgments: We thank to Baveesh Pudhuvai (BC CAS, Budweis, Czech Republic) for help in the PCR verification of *Buddenbrockia* stefin. We also thank Ivan Fiala (BC CAS, Budweis, Czech Republic) for providing suggestions to improve the manuscript and for sharing *M. lieberkuehni* and *N. pickii* transcriptomic data. We are grateful to Hanna Hartikainen (ETH Zurich, Switzerland) for sharing *T. bryosalmonae* genome data.

Conflicts of Interest: The authors declare no conflict of interest. The funders had no role in the design of the study; in the collection, analyses, or interpretation of data; in the writing of the manuscript; or in the decision to publish the results.

References

1. Cancela, M.; Corvo, I.; da Silva, E.; Teichmann, A.; Roche, L.; Diaz, A.; Tort, J.F.; Ferreira, H.B.; Zaha, A. Functional characterization of single-domain cystatin-like cysteine proteinase inhibitors expressed by the trematode *Fasciola hepatica*. *Parasitology* **2017**, *144*, 1695–1707. [CrossRef] [PubMed]
2. Buša, M.; Matoušková, Z.; Řezáčová, P.; Bartošová-Sojková, P.; Štefanič, S.; Mareš, M. Evolutionary upgrade of stefins for secretion in parasites. In Proceedings of the 11th General Meeting of the International Proteolysis Society "Interfaces in Proteolysis", Mariánské Lázně, Czech Republic, 29 September–4 October 2019.

3. Siricoon, S.; Grams, S.V.; Grams, R. Efficient inhibition of cathepsin B by a secreted type 1 cystatin of *Fasciola gigantica*. *Mol. Biochem. Parasit.* **2012**, *186*, 126–133. [CrossRef] [PubMed]
4. Tarasuk, M.; Grams, S.V.; Viyanant, V.; Grams, R. Type I cystatin (stefin) is a major component of *Fasciola gigantica* excretion/secretion product. *Mol. Biochem. Parasit.* **2009**, *167*, 60–71. [CrossRef] [PubMed]
5. Abrahamson, M.; Alvarez-Fernandez, M.; Nathanson, C. Cystatins. *Biochem. Soc. Symp.* **2003**, *70*, 179–199.
6. Turk, V.; Stoka, V.; Turk, D. Cystatins: Biochemical and structural properties, and medical relevance. *Front. Biosci.* **2008**, *13*, 5406–5420. [CrossRef]
7. Turk, V.; Turk, D.; Dolenc, I.; Stoka, V. Characteristics, structure, and biological role of stefins (type-1 cystatins) of human, other mammals, and parasite origin. *Acta Chim. Slov.* **2019**, *66*, 5–17. [CrossRef]
8. Ranasinghe, S.; McManus, D. Protease inhibitors of parasitic flukes: Emerging roles in parasite survival and immune defence. *Trends Parasitol.* **2017**, *33*, 400–413. [CrossRef]
9. Chmelař, J.; Kotál, J.; Langhansová, H.; Kotsyfakis, M. Protease inhibitors in tick saliva: The role of serpins and cystatins in tick-host-pathogen interaction. *Front. Cell. Infect. Microbiol.* **2017**, *7*, 216. [CrossRef]
10. Gregory, W.F.; Maizels, R.M. Cystatins from filarial parasites: Evolution, adaptation and function in the host-parasite relationship. *Int. J. Biochem. Cell Biol.* **2008**, *40*, 1389–1398. [CrossRef]
11. Wang, Y.J.; Wu, L.Y.; Liu, X.C.; Wang, S.; Ehsan, M.; Yan, R.F.; Song, X.K.; Xu, L.X.; Li, X.R. Characterization of a secreted cystatin of the parasitic nematode *Haemonchus contortus* and its immune-modulatory effect on goat monocytes. *Parasit. Vectors* **2017**, *10*, 425. [CrossRef]
12. Kotál, J.; Stergiou, N.; Buša, M.; Chlastáková, A.; Beránková, Z.; Řezáčová, P.; Langhansová, H.; Schwarz, A.; Calvo, E.; Kopecký, J.; et al. The structure and function of Iristatin, a novel immunosuppressive tick salivary cystatin. *Cell Mol. Life Sci.* **2019**, *76*, 2003–2013. [CrossRef] [PubMed]
13. Schwarz, A.; Valdés, J.J.; Kotsyfakis, M. The role of cystatins in tick physiology and blood feeding. *Ticks Tick Borne Dis.* **2012**, *3*, 117–127. [CrossRef] [PubMed]
14. Klotz, C.; Ziegler, T.; Figueiredo, A.S.; Rausch, S.; Hepworth, M.R.; Obsivac, N.; Sers, C.; Lang, R.; Hammerstein, P.; Lucius, R.; et al. A helminth immunomodulator exploits host signaling events to regulate cytokine production in macrophages. *PLoS Pathog.* **2011**, *7*, e1001248. [CrossRef] [PubMed]
15. Maizels, R.M.; Gomez-Escobar, N.; Gregory, W.F.; Murray, J.; Zang, X.X. Immune evasion genes from filarial nematodes. *Int. J. Parasitol.* **2001**, *31*, 889–898. [CrossRef]
16. Klotz, C.; Ziegler, T.; Daniłowicz-Luebert, E.; Hartmann, S. Cystatins of parasitic organisms. *Adv. Exp. Med. Biol.* **2011**, *712*, 208–221. [PubMed]
17. Khatri, V.; Chauhan, N.; Kalyanasundaram, R. Parasite cystatin: Immunomodulatory molecule with therapeutic activity against immune mediated disorders. *Pathogens* **2020**, *9*, 431. [CrossRef] [PubMed]
18. Kopitar-Jerala, N. The role of cystatins in cells of the immune system. *FEBS Lett.* **2006**, *580*, 6295–6301. [CrossRef]
19. Salát, J.; Paesen, G.C.; Rezáčová, P.; Kotsyfakis, M.; Kovářová, Z.; Šanda, M.; Majtán, J.; Grunclová, L.; Horká, H.; Andersen, J.F.; et al. Crystal structure and functional characterization of an immunomodulatory salivary cystatin from the soft tick *Ornithodoros moubata*. *Biochem. J.* **2010**, *429*, 103–112. [CrossRef]
20. Hartmann, S.; Lucius, R. Modulation of host immune responses by nematode cystatins. *Int. J. Parasitol.* **2003**, *33*, 1291–1302. [CrossRef]
21. Schierack, P.; Lucius, R.; Sonnenburg, B.; Schilling, K.; Hartmann, S. Parasite-specific immunomodulatory functions of filarial cystatin. *Infect. Immun.* **2003**, *71*, 2422–2429. [CrossRef]
22. Khaznadji, E.; Collins, P.; Dalton, J.P.; Bigot, Y.; Moire, N. A new multi-domain member of the cystatin superfamily expressed by *Fasciola hepatica*. *Int. J. Parasitol.* **2005**, *35*, 1115–1125. [CrossRef] [PubMed]
23. McKerrow, J.; Engel, J.; Caffrey, C. Cysteine protease inhibitors as chemotherapy for parasitic infections. *Bioorg. Med. Chem.* **1999**, *7*, 639–644. [CrossRef]
24. Tang, B.; Liu, M.; Wang, L.; Yu, S.; Shi, H.; Boireau, P.; Cozma, V.; Wu, X.; Liu, X. Characterisation of a high-frequency gene encoding a strongly antigenic cystatin-like protein from *Trichinella spiralis* at its early invasion stage. *Parasit. Vectors* **2015**, *8*, 78. [CrossRef] [PubMed]
25. Abdulla, M.; Lim, K.; Sajid, M.; McKerrow, J.; Caffrey, C. Schistosomiasis mansoni: Novel chemotherapy using a cysteine protease inhibitor. *PLoS Med.* **2007**, *4*, 130–138. [CrossRef] [PubMed]
26. Das, P.; Alam, M.; Paik, D.; Karmakar, K.; De, T.; Chakrahorti, T. Protease inhibitors in potential drug development for leishmaniasis. *Indian J. Biochem. Biol.* **2013**, *50*, 363–376.
27. Duschak, V. Targets and patented drugs for chemotherapy of Chagas disease in the last 15 years-period. *Recent Pat. Antiinfect. Drug Discov.* **2016**, *11*, 74–173. [CrossRef]
28. Smallwood, T.B.; Giacomin, P.R.; Loukas, A.; Mulvenna, J.P.; Clark, R.J.; Miles, J.J. Helminth immunomodulation in autoimmune disease. *Front. Immunol.* **2017**, *8*, 453. [CrossRef]
29. Breznik, B.; Mitrovic, A.; Lah, T.T.; Kos, J. Cystatins in cancer progression: More than just cathepsin inhibitors. *Biochimie* **2019**, *166*, 233–250. [CrossRef]
30. Kordiš, D.; Turk, V. Phylogenomic analysis of the cystatin superfamily in eukaryotes and prokaryotes. *BMC Evol. Biol.* **2009**, *9*, 266. [CrossRef]

31. Martinez, M.; Diaz, I. The origin and evolution of plant cystatins and their target cysteine proteinases indicate a complex functional relationship. *BMC Evol. Biol.* **2008**, *8*, 198. [CrossRef]
32. Brown, W.M.; Dziegielewska, K.M. Friends and relations of the cystatin superfamily—New members and their evolution. *Protein Sci.* **1997**, *6*, 5–12. [CrossRef] [PubMed]
33. Rawlings, N.D.; Barrett, A.J. Evolution of proteins of the cystatin superfamily. *J. Mol. Evol.* **1990**, *30*, 60–71. [CrossRef] [PubMed]
34. Verdes, A.; Simpson, D.; Holford, M. Are fireworms venomous? Evidence for the convergent evolution of toxin homologs in three species of fireworms (Annelida, Amphinomidae). *Genome Biol. Evol.* **2018**, *10*, 249–268. [CrossRef] [PubMed]
35. Cuesta-Astroz, Y.; Scholte, L.L.S.; Pais, F.S.M.; Oliveira, G.; Nahum, L.A. Evolutionary analysis of the cystatin family in three *Schistosoma* species. *Front. Genet.* **2014**, *5*, 206. [CrossRef] [PubMed]
36. Guo, A.J. Comparative analysis of cystatin superfamily in Platyhelminths. *PLoS ONE* **2015**, *10*, e0124683. [CrossRef]
37. Ilgová, J.; Jedličková, L.; Dvořaková, H.; Benovics, M.; Mikeš, L.; Janda, L.; Vorel, J.; Roudnický, P.; Potěšil, D.; Zdráhal, Z.; et al. A novel type I cystatin of parasite origin with atypical legumain-binding domain. *Sci. Rep.* **2017**, *7*, 17526. [CrossRef]
38. Dong, X.W.; Xu, J.; Song, H.Y.; Liu, Y.C.; Wu, M.D.; Zhang, H.J.; Jing, B.; Lai, W.M.; Gu, X.B.; Xie, Y.; et al. Molecular characterization of a *Dirofilaria immitis* cysteine protease inhibitor (cystatin) and its possible role in filarial immune evasion. *Genes* **2019**, *10*, 300. [CrossRef]
39. Santamaria, M.E.; Hernandez-Crespo, P.; Ortego, F.; Grbic, V.; Grbic, M.; Diaz, I.; Martinez, M. Cysteine peptidases and their inhibitors in *Tetranychus urticae*: A comparative genomic approach. *BMC Genom.* **2012**, *13*, 307. [CrossRef]
40. Rangel, C.K.; Parizi, L.F.; Sabadin, G.A.; Costa, E.P.; Romeiro, N.C.; Isezaki, M.; Githaka, N.W.; Seixas, A.; Logullo, C.; Konnai, S.; et al. Molecular and structural characterization of novel cystatins from the taiga tick *Ixodes persulcatus*. *Ticks Tick Borne Dis.* **2017**, *8*, 432–441. [CrossRef]
41. Ibelli, A.M.G.; Hermance, M.M.; Kim, T.K.; Gonzalez, C.L.; Mulenga, A. Bioinformatics and expression analyses of the *Ixodes scapularis* tick cystatin family. *Exp. Appl. Acarol.* **2013**, *60*, 41–53. [CrossRef]
42. Alonso, J.; Martinez, M. Insights into the molecular evolution of peptidase inhibitors in arthropods. *PLoS ONE* **2017**, *12*, e0187643. [CrossRef] [PubMed]
43. Weinstein, S.B.; Kuris, A.M. Independent origins of parasitism in Animalia. *Biol. Lett.* **2016**, *12*, 20160324. [CrossRef] [PubMed]
44. Dnyansagar, R.; Zimmermann, B.; Moran, Y.; Praher, D.; Sundberg, P.; Moller, L.; Technau, U. Dispersal and speciation: The cross Atlantic relationship of two parasitic cnidarians. *Mol. Phylogenet. Evol.* **2018**, *126*, 346–355. [CrossRef] [PubMed]
45. Stefanik, D.J.; Lubinski, T.J.; Granger, B.R.; Byrd, A.L.; Reitzel, A.M.; DeFilippo, L.; Lorenc, A.; Finnerty, J.R. Production of a reference transcriptome and transcriptomic database (EdwardsiellaBase) for the lined sea anemone, *Edwardsiella lineata*, a parasitic cnidarian. *BMC Genom.* **2014**, *15*, 71. [CrossRef]
46. Shpirer, E.; Chang, E.S.; Diamant, A.; Rubinstein, N.; Cartwright, P.; Huchon, D. Diversity and evolution of myxozoan minicollagens and nematogalectins. *BMC Evol. Biol.* **2014**, *14*, 205. [CrossRef]
47. Chang, E.S.; Neuhof, M.; Rubinstein, N.D.; Diamant, A.; Philippe, H.; Huchon, D.; Cartwright, P. Genomic insights into the evolutionary origin of Myxozoa within Cnidaria. *Proc. Natl. Acad. Sci. USA* **2015**, *112*, 14912–14917. [CrossRef]
48. Kumar, G.; Abd-Elfattah, A.; El-Matbouli, M. Identification of differentially expressed genes of brown trout (*Salmo trutta*) and rainbow trout (*Oncorhynchus mykiss*) in response to *Tetracapsuloides bryosalmonae* (Myxozoa). *Parasitol. Res.* **2015**, *114*, 929–939. [CrossRef]
49. Jiménez-Guri, E.; Philippe, H.; Okamura, B.; Holland, P.W.H. *Buddenbrockia* is a cnidarian worm. *Science* **2007**, *317*, 116–118. [CrossRef]
50. Alama-Bermejo, G.; Meyer, E.; Atkinson, S.D.; Holzer, A.S.; Wiśniewska, M.M.; Kolísko, M.; Bartholomew, J.L. Transcriptome-wide comparisons and virulence gene polymorphisms of host-associated genotypes of the cnidarian parasite *Ceratonova shasta* in salmonids. *Genome Biol. Evol.* **2020**, *12*, 1258–1276. [CrossRef]
51. Nesnidal, M.; Helmkampf, M.; Bruchhaus, I.; El-Matbouli, M.; Hausdorf, B. Agent of whirling disease meets orphan worm: Phylogenomic analyses firmly place Myxozoa in Cnidaria. *PLoS ONE* **2013**, *8*, e54576. [CrossRef]
52. Foox, J.; Ringuette, M.; Desser, S.S.; Siddall, M.E. In silico hybridization enables transcriptomic illumination of the nature and evolution of Myxozoa. *BMC Genom.* **2015**, *16*, 840. [CrossRef] [PubMed]
53. Yang, Y.L.; Xiong, J.; Zhou, Z.G.; Huo, F.M.; Miao, W.; Ran, C.; Liu, Y.C.; Zhang, J.Y.; Feng, J.M.; Wang, M.; et al. The genome of the myxosporean *Thelohanellus kitauei* shows adaptations to nutrient acquisition within its fish host. *Genome Biol. Evol.* **2014**, *6*, 3182–3198. [CrossRef] [PubMed]
54. Hartigan, A.; Kosakyan, A.; Pecková, H.; Eszterbauer, E.; Holzer, A.S. Transcriptome of *Sphaerospora molnari* (Cnidaria, Myxosporea) blood stages provides proteolytic arsenal as potential therapeutic targets against sphaerosporosis in common carp. *BMC Genom.* **2020**, *21*, 404. [CrossRef] [PubMed]
55. Evans, N.M.; Lindner, A.; Raikova, E.V.; Collins, A.G.; Cartwright, P. Phylogenetic placement of the enigmatic parasite, *Polypodium hydriforme*, within the Phylum Cnidaria. *BMC Evol. Biol.* **2008**, *8*, 139. [CrossRef] [PubMed]
56. Okamura, B.; Gruhl, A. Myxozoan affinities and route to endoparasitism. In *Myxozoan Evolution, Ecology and Development*; Okamura, B., Gruhl, A., Bartholomew, J.L., Eds.; Springer: Cham, Switzerland, 2015; pp. 23–44.
57. Fiala, I.; Bartošová-Sojková, P.; Okamura, B.; Hartikainen, H. Adaptive radiation and evolution within the Myxozoa. In *Myxozoan Evolution, Ecology and Development*; Okamura, B., Gruhl, A., Bartholomew, J.L., Eds.; Springer: Cham, Switzerland, 2015; pp. 69–84.
58. Feng, J.M.; Xiong, J.; Zhang, J.Y.; Yang, Y.L.; Yao, B.; Zhou, Z.G.; Miao, W. New phylogenomic and comparative analyses provide corroborating evidence that Myxozoa is Cnidaria. *Mol. Phylogenet. Evol.* **2014**, *81*, 10–18. [CrossRef]

59. Stubbs, M.; Laber, B.; Bode, W.; Huber, R.; Jerala, R.; Lenarcic, B.; Turk, V. The refined 2.4A X-ray crystal structure of recombinant human stefin B in complex with the cysteine proteinase papain: A novel type of proteinase inhibitor interaction. *EMBO J.* **1990**, *9*, 1939–1947. [CrossRef]

60. Rawlings, N.D.; Alan, J.; Thomas, P.D.; Huang, X.D.; Bateman, A.; Finn, R.D. The MEROPS database of proteolytic enzymes, their substrates and inhibitors in 2017 and a comparison with peptidases in the PANTHER database. *Nucleic Acids Res.* **2018**, *46*, D624–D632. [CrossRef]

61. Potter, S.C.; Luciani, A.; Eddy, S.R.; Park, Y.; Lopez, R.; Finn, R.D. HMMER web server: 2018 update. *Nucleic Acids Res.* **2018**, *46*, W200–W204. [CrossRef]

62. Kearse, M.; Moir, R.; Wilson, A.; Stones-Havas, S.; Cheung, M.; Sturrock, S.; Buxton, S.; Cooper, A.; Markowitz, S.; Duran, C. Geneious Basic: An integrated and extendable desktop software platform for the organization and analysis of sequence data. *Bioinformatics* **2012**, *28*, 1647–1649. [CrossRef] [PubMed]

63. Stamatakis, A. RAxML version 8: A tool for phylogenetic analysis and post-analysis of large phylogenies. *Bioinformatics* **2014**, *30*, 1312–1313. [CrossRef]

64. Trifinopoulos, J.; Nguyen, L.T.; von Haeseler, A.; Minh, B.Q. W-IQ-TREE: A fast online phylogenetic tool for maximum likelihood analysis. *Nucleic Acids Res.* **2016**, *44*, W232–W235. [CrossRef] [PubMed]

65. Ronquist, F.; Teslenko, M.; van der Mark, P.; Ayres, D.L.; Darling, A.; Höhna, S.; Larget, B.; Liu, L.; Suchard, M.A.; Huelsenbeck, J.P. MrBayes 3.2: Efficient Bayesian phylogenetic inference and model choice across a large model space. *Syst. Biol.* **2012**, *61*, 539–542. [CrossRef] [PubMed]

66. Ceroni, A.; Passerini, A.; Vullo, A.; Frasconi, P. DISULFIND: A disulfide bonding state and cysteine connectivity prediction server. *Nucleic Acids Res.* **2006**, *34*, W177–W181. [CrossRef] [PubMed]

67. Almagro Armenteros, J.J.; Tsirigos, K.D.; Sønderby, C.K.; Petersen, T.N.; Winther, O.; Brunak, S.; von Heijne, G.; Nielsen, H. SignalP 5.0 improves signal peptide predictions using deep neural networks. *Nat. Biotechnol.* **2019**, *37*, 420–423. [CrossRef] [PubMed]

68. Crooks, G.E.; Hon, G.; Chandonia, J.M.; Brenner, S.E. WebLogo: A sequence logo generator. *Genome Res.* **2004**, *14*, 1188–1190. [CrossRef] [PubMed]

69. Mao, Y.F.; Satoh, N. A likely ancient genome duplication in the speciose reef-building coral genus, *Acropora*. *Iscience* **2019**, *13*, 20–32. [CrossRef] [PubMed]

70. Kamm, K.; Osigus, H.J.; Stadler, P.F.; DeSalle, R.; Schierwater, B. *Trichoplax* genomes reveal profound admixture and suggest stable wild populations without bisexual reproduction. *Sci. Rep.* **2018**, *8*, 11168. [CrossRef] [PubMed]

71. Eitel, M.; Francis, W.; Varoqueaux, F.; Daraspe, J.; Osigus, H.; Krebs, S.; Vargas, S.; Blum, H.; Williams, G.; Schierwater, B.; et al. Comparative genomics and the nature of placozoan species. *PLoS Biol.* **2018**, *16*, e3000032. [CrossRef]

72. Srivastava, M.; Begovic, E.; Chapman, J.; Putnam, N.; Hellsten, U.; Kawashima, T.; Kuo, A.; Mitros, T.; Salamov, A.; Carpenter, M.; et al. The *Trichoplax* genome and the nature of placozoans. *Nature* **2008**, *454*, 955–960. [CrossRef]

73. Lenarcic, B.; Ritonja, A.; Strukelj, B.; Turk, B.; Turk, V. Equistatin, a new inhibitor of cysteine proteinases from *Actinia equina*, is structurally related to thyroglobulin type-1 domain. *J. Biol. Chem.* **1997**, *272*, 13899–13903. [CrossRef]

74. Irving, J.A.; Pike, R.N.; Dai, W.W.; Bromme, D.; Worrall, D.M.; Silverman, G.A.; Coetzer, T.H.T.; Dennison, C.; Bottomley, S.P.; Whisstock, J.C. Evidence that serpin architecture intrinsically supports papain-like cysteine protease inhibition: Engineering alpha(1)-antitrypsin to inhibit cathepsin proteases. *Biochemistry* **2002**, *41*, 4998–5004. [CrossRef] [PubMed]

75. Eszterbauer, E.; Sipos, D.; Kaján, G.L.; Szegö, D.; Fiala, I.; Holzer, A.S.; Bartošová-Sojková, P. Genetic diversity of serine protease inhibitors in myxozoan (Cnidaria, Myxozoa) fish parasites. *Microorganisms* **2020**, *8*, 1502. [CrossRef] [PubMed]

76. Geadkaew, A.; Kosa, N.; Siricoon, S.; Grams, S.V.; Grams, R. A 170 kDa multi-domain cystatin of *Fasciola gigantica* is active in the male reproductive system. *Mol. Biochem. Parasit.* **2014**, *196*, 100–107. [CrossRef] [PubMed]

77. Pol, E.; Bjork, I. Importance of the second binding loop and the C-terminal end of cystatin B (stefin B) for inhibition of cysteine proteinases. *Biochemistry* **1999**, *38*, 10519–10526. [CrossRef] [PubMed]

78. Jerala, R.; Kroon-Zitko, L.; Kopitar, M.; Popovic, T.; Turk, V. Deletion of the carboxy terminal part of stefin B does not have a major effect for binding to papain. *Biomed. Biochim. Acta* **1991**, *50*, 627–629. [PubMed]

79. Hewitson, J.P.; Grainger, J.R.; Maizels, R.M. Helminth immunoregulation: The role of parasite secreted proteins in modulating host immunity. *Mol. Biochem. Parasit.* **2009**, *167*, 1–11. [CrossRef]

80. Vendelova, E.; de Lima, J.C.; Lorenzatto, K.R.; Monteiro, K.M.; Mueller, T.; Veepaschit, J.; Grimm, C.; Brehm, K.; Hrčková, G.; Lutz, M.B.; et al. Proteomic analysis of excretory-secretory products of *Mesocestoides corti* metacestodes reveals potential suppressors of dendritic cell functions. *PLoS Negl. Trop. Dis.* **2016**, *10*, e0005061. [CrossRef]

81. Di Maggio, L.S.; Tirloni, L.; Pinto, A.F.M.; Diedrich, J.K.; Yates, J.R.; Benavides, U.; Carmona, C.; Vaz, I.D.; Berasain, P. Across intra-mammalian stages of the liver fluke *Fasciola hepatica*: A proteomic study. *Sci. Rep.* **2016**, *6*, 32796. [CrossRef]

82. Kang, J.M.; Lee, K.H.; Sohn, W.M.; Na, B.K. Identification and functional characterization of CsStefin-1, a cysteine protease inhibitor of *Clonorchis sinensis*. *Mol. Biochem. Parasit.* **2011**, *177*, 126–134. [CrossRef]

83. Cantacessi, C.; Mulvenna, J.; Young, N.D.; Kasny, M.; Horak, P.; Aziz, A.; Hofmann, A.; Loukas, A.; Gasser, R.B. A deep exploration of the transcriptome and "excretory/secretory" proteome of adult *Fascioloides magna*. *Mol. Cell. Proteomics* **2012**, *11*, 1340–1353. [CrossRef]

84. Zhang, F.; Yang, Y.; Gao, C.; Yao, Y.; Xia, R.; Hu, J.; Ran, C.; Zhang, Z.; Zhou, Z. Bioinformatics analysis and characterization of a secretory cystatin from *Thelohanellus kitauei*. *AMB Express* **2020**, *10*, 116. [CrossRef] [PubMed]

85. Korytář, T.; Wiegertjes, G.F.; Zusková, E.; Tomanová, A.; Lisnerová, M.; Patra, S.; Sieranski, V.; Šíma, R.; Born-Torrijos, A.; Wentzel, A.S.; et al. The kinetics of cellular and humoral immune responses of common carp to presporogonic development of the myxozoan *Sphaerospora molnari*. *Parasit. Vectors* **2019**, *12*, 208. [CrossRef] [PubMed]

86. Bjork, S.J.; Zhang, Y.A.; Hurst, C.N.; Alonso-Naveiro, M.E.; Alexander, J.D.; Sunyer, J.O.; Bartholomew, J.L. Defenses of susceptible and resistant Chinook salmon (*Onchorhynchus tshawytscha*) against the myxozoan parasite *Ceratomyxa shasta*. *Fish Shellfish Immun.* **2014**, *37*, 87–95. [CrossRef] [PubMed]

87. Bailey, C.; Strepparava, N.; Wahli, T.; Segner, H. Exploring the immune response, tolerance and resistance in proliferative kidney disease of salmonids. *Dev. Comp. Immunol.* **2019**, *90*, 165–175. [CrossRef] [PubMed]

88. Piazzon, M.C.; Estensoro, I.; Calduch-Giner, J.A.; del Pozo, R.; Picard-Sánchez, A.; Perez-Sánchez, J.; Sitjà-Bobadilla, A. Hints on T cell responses in a fish-parasite model: *Enteromyxum leei* induces differential expression of T cell signature molecules depending on the organ and the infection status. *Parasit. Vectors* **2018**, *11*, 443. [CrossRef] [PubMed]

89. Pérez-Cordón, G.; Estensoro, I.; Benedito-Palos, L.; Calduch-Giner, J.A.; Sitjà-Bobadilla, A.; Pérez-Sanchez, J. Interleukin gene expression is strongly modulated at the local level in a fish-parasite model. *Fish Shellfish Immun.* **2014**, *37*, 201–208. [CrossRef]

90. Gorgoglione, B.; Wang, T.; Secombes, C.J.; Holland, J.W. Immune gene expression profiling of proliferative kidney disease in rainbow trout *Oncorhynchus mykiss* reveals a dominance of anti-inflammatory, antibody and Th cell-like activities. *Fish Shellfish Immun.* **2013**, *34*, 1654. [CrossRef]

91. Sommerset, I.; Krossoy, B.; Biering, E.; Frost, P. Vaccines for fish in aquaculture. *Expert Rev. Vaccines* **2005**, *4*, 89–101. [CrossRef] [PubMed]

92. Sitjà-Bobadilla, A.; Schmidt-Posthaus, H.; Wahli, T.; Holland, J.W.; Secombes, C.J. Fish immune responses to Myxozoa. In *Myxozoan Evolution, Ecology and Development*; Okamura, B., Gruhl, A., Bartholomew, J.L., Eds.; Springer: Cham, Switzerland, 2015; pp. 253–280.

93. Rogers, R.L.; Hartl, D.L. Chimeric genes as a source of rapid evolution in *Drosophila melanogaster*. *Mol. Biol. Evol.* **2012**, *29*, 517–529. [CrossRef] [PubMed]

94. Rogers, R.L.; Bedford, T.; Hartl, D.L. Formation and longevity of chimeric and duplicate genes in *Drosophila melanogaster*. *Genetics* **2009**, *181*, 313–322. [CrossRef] [PubMed]

95. Eszterbauer, E.; Atkinson, S.; Diamant, A.; Morris, D.J.; El-Matbouli, M.; Hartikainen, H. Myxozoan life cycles: Practical approaches and insights. In *Myxozoan Evolution, Ecology and Development*; Okamura, B., Gruhl, A., Bartholomew, J.L., Eds.; Springer: Cham, Switzerland, 2015; pp. 175–198.

96. Holzer, A.S.; Bartošová-Sojková, P.; Born-Torrijos, A.; Lövy, A.; Hartigan, A.; Fiala, I. The joint evolution of the Myxozoa and their alternate hosts: A cnidarian recipe for success and vast biodiversity. *Mol. Ecol.* **2018**, *27*, 1651–1666. [CrossRef] [PubMed]

97. Melillo, D.; Marino, R.; Italiani, P.; Boraschi, D. Innate immune memory in invertebrate metazoans: A critical appraisal. *Front. Immunol.* **2018**, *9*, 1915. [CrossRef] [PubMed]

98. Dickinson, D.P. Salivary (SD-type) cystatins: Over one billion years in the making-but to what purpose? *Crit. Rev. Oral Biol. Med.* **2002**, *13*, 485–508. [CrossRef] [PubMed]

biology

Article

Ranacyclin-NF, a Novel Bowman–Birk Type Protease Inhibitor from the Skin Secretion of the East Asian Frog, *Pelophylax nigromaculatus*

Tao Wang [†], Yangyang Jiang [†], Xiaoling Chen *, Lei Wang, Chengbang Ma, Xinping Xi, Yingqi Zhang, Tianbao Chen, Chris Shaw and Mei Zhou

Natural Drug Discovery Group, School of Pharmacy, Queen's University Belfast, Belfast BT9 7BL, UK; twang13@qub.ac.uk (T.W.); yjiang12@qub.ac.uk (Y.J.); l.wang@qub.ac.uk (L.W.); c.ma@qub.ac.uk (C.M.); x.xi@qub.ac.uk (X.X.); zhangyingqi08@sina.com (Y.Z.); t.chen@qub.ac.uk (T.C.); chris.shaw@qub.ac.uk (C.S.); m.zhou@qub.ac.uk (M.Z.)
* Correspondence: x.chen@qub.ac.uk; Tel.: +44-28-9097-2200
† These authors contributed equally to this work.

Received: 7 May 2020; Accepted: 29 June 2020; Published: 2 July 2020

Abstract: Serine protease inhibitors are found in plants, animals and microorganisms, where they play important roles in many physiological and pathological processes. Inhibitor scaffolds based on natural proteins and peptides have gradually become the focus of current research as they tend to bind to their targets with greater specificity than small molecules. In this report, a novel Bowman–Birk type inhibitor, named ranacyclin-NF (RNF), is described and was identified in the skin secretion of the East Asian frog, *Pelophylax nigromaculatus*. A synthetic replicate of the peptide was subjected to a series of functional assays. It displayed trypsin inhibitory activity with an inhibitory constant, Ki, of 447 nM and had negligible direct cytotoxicity. No observable direct antimicrobial activity was found but RNF improved the therapeutic potency of Gentamicin against Methicillin-resistant *Staphylococcus aureus* (MRSA). RNF shared significant sequence similarity to previously reported and related inhibitors from *Odorrana grahami* (ORB) and *Rana esculenta* (ranacyclin-T), both of which were found to be multi-functional. Two analogues of RNF, named ranacyclin-NF1 (RNF1) and ranacyclin-NF3L (RNF3L), were designed based on some features of ORB and ranacyclin-T to study structure–activity relationships. Structure–activity studies demonstrated that residues outside of the trypsin inhibitory loop (TIL) may be related to the efficacy of trypsin inhibitory activity.

Keywords: Bowman–Birk inhibitor; ranacyclin; trypsin inhibitor; structure–activity relationship; synergistic effect; Gentamicin

1. Introduction

Serine proteases exist in almost all living organisms [1], and, in humans, they are intimately involved in many disease processes, such as in infections, coagulation disorders, cancer and inflammation [2,3]. Although most current serine protease-targeted drugs are small molecules, there is an increasing attention shift to polypeptide inhibitors as these more complex molecules offer an achievable approach for optimising the affinity and specificity of inhibitors. Inhibitor scaffolds are normally contained in natural proteins or peptides which have been optimised and improved through a process of parallel evolution for activities against endogenous targets [4].

Bowman–Birk inhibitors (BBIs) exist in a wide range of plants including sunflowers, soybeans, legumes and so on [5]. They are classical serine proteinase inhibitors that contain a highly-conserved disulphide loop structure consisting of nine residues ($CTP_1SXPPXC$) [6–8]. BBI peptides derived from amphibians are slightly different and typically contain a disulphide-bridged hendecapeptide loop (CWTKSXPPXPC), containing 11 residues rather than nine residues [9,10]. For all natural trypsin-inhibition domains, the residue at the P1 position is lysine or arginine, whereas the P1 residue is typically leucine or tyrosine in chymotrypsin inhibitory domains [11]. BBIs have been extensively studied and found to have a wide range of functions, such as anti-inflammatory, anticancer and anti-bacterial. In addition, compared to most antimicrobial peptides (AMPs), BBIs are safe orally-active peptides whose therapeutic effects have been proven in animal models [12]. A thorough understanding of BBIs and their therapeutic potential could help in their further development as clinical agents.

The term "ranacyclins" was first used in 2003 to describe two cyclic AMPs, ranacyclin-T and ranacyclin-E, isolated from the skin secretion of *Rana temporaria* and *Rana esculenta*, respectively [13]. According to the protein database InterPro (https://www.ebi.ac.uk/interpro/), currently, there are 73 ranacyclin-type peptides have been identified from the frog skin secretion. The residue length of ranacyclins ranges from 13 to 30, and most are composed of 17 amino acid residues. They share a similar peptide sequence and contain a highly conserved mutated Bowman–Birk trypsin inhibitory loop (TIL). Although ranacyclins share similar structural features, their bioactivities are extremely diversified [14]. pLR from *Lithobates pipiens* and pYR from *Rana sevosa*, for example, were reported with immunomodulatory properties and could suppress the early development of granulocyte macrophage colonies from bone marrow stem cells [15,16]. ZDPI from *Amolops loloensis* could inhibit the aggregation of platelet [17]. Ranacyclin-HB1 from *Pelophylax hubeiensis* showed obvious antioxidant activity and could scavenge the free radicals rapidly [18]. These diversified functions of ranacyclins might be associated with the species adaption to different environments [19,20].

Compared with most AMPs, peptides in ranacyclin with multiple functions were considered as a more promising candidate for a novel generation of antibiotic agents than most AMPs due to their excellent hydrolysis-tolerance capacity and low cytotoxicity [21]. However, compared with common AMP families, such as dermaseptins or brevinins, reports in these families are relatively less. Therefore, studies and modification on ranacyclins would favour to a better understanding of this family and put them into an application. In this study, a novel Bowman–Birk type ranacyclin, named ranacyclin-NF (RNF), was discovered in the skin secretion of *Pelophylax nigromaculatus* and was subsequently structurally- and functionally-characterised. To study the structure–activity relationships of this ranacyclin in more depth and based on previous publications, two analogues, named ranacyclin-NF1 (RNF1) and ranacyclin-NF3L (RNF3L), were designed and subjected to bioactivity assessment in parallel with a synthetic replicate of the natural peptide.

2. Materials and Methods

2.1. Acquisition of Skin Secretion

Four East Asian Frogs, *Pelophylax nigromaculatus*, were captured in China. All frogs were mature adults and their skin secretions were obtained from the dorsal skin by gentle transdermal electrical stimulation. The skin secretion was washed from the skin with de-ionised water (ddH$_2$O), collected within one 50-mL tube, frozen in liquid nitrogen, lyophilised and stored at -20 °C prior to analysis [13]. The study was performed according to the guidelines of the UK Animal (Scientific Procedures) Act 1986, under project license PPL 2694 (M.Z.), issued by the Department of Health, Social Services and Public Safety, Northern Ireland. Procedures have been vetted by the Institutional Animal Care and Use Committee (IACUC) of Queen's University Belfast and approved on 1 March, 2011.

2.2. "Shotgun" Cloning and Protein Bioinformatic Analyses

Polyadenylated mRNA was removed from the skin secretion of *Pelophylax nigromaculatus* using a Dynabeads mRNA Direct Kit (Thermo Fisher Scientific Inc., Waltham, MA, USA). The first cDNA was then obtained by reverse transcription and the cDNA library was constructed by a RACE technique using an Advantage^{TM} 2 PCR Kit (BD Clontech, Oxford, UK). Then, the full-length sequence of the mRNA transcript encoding the RNF precursor was captured using a SMART-RACE kit (Clontech

U.K.). The degenerate primer was 5′-CCCRAAKATGTTSACCTYRAAGAAA-3′, designed from a highly-conserved domain within the 5′- untranslated regions of closely-related *Rana* species. The RACE products were then cloned by use of a pGEM T-easy vector system (Promega, Southampton, UK) and sequenced by an ABI 3100 automated sequencer (Applied Biosystems, Foster City, CA, USA). The target cDNA and corresponding translated protein sequence were compared to homologous entries in the GenBank public database with BLASTn and BLASTp, respectively (National Centre for Biotechnology Information (NCBI), http://www.ncbi.nlm.nih.gov). The homologous protein sequences mainly came from different species of frog and these sequences were then aligned with the precursor of the novel peptide. According to the results of the structural bioinformatic analyses, the mature peptide sequence was predicted [22].

2.3. Solid-Phase Peptide Synthesis and Confirmation of Structure

RNF and its analogues, RNF1 and RNF3L, were synthesised using standard Fmoc amino acids on a Tribute automated peptide synthesiser (Protein Technologies Inc., Tucson, AZ, USA). When the synthesis was completed, the peptides were cleaved from the resin and de-protected. Using a weak solution of hydrogen peroxide, the samples were oxidised to form the disulphide loop structure. After purification on an Adept CECIL4200 RP-HPLC (Amersham Biosciences Inc., Piscatawa, NJ, USA), Column: Phenomenex Aeris Peptide, C18, 250 mm × 10.0 mm (Phenomenex, Macclesfeld, UK), the masses of peptide samples were confirmed by using matrix-assisted laser desorption/ionisation time-of-flight (MALDI-TOF) mass spectrometry (Perseptive Biosystems, Framingham, MA, USA). Alpha-cyano-4-hydroxycinnamic acid (CHCA) (Sigma Chemical Co., St. Louis, MO, USA) was used as matrix and was dissolved in TFA/H_2O/acetonitrile (0.05/49.95/50, $v/v/v$).

2.4. Circular Dichroism (CD) Analysis

The secondary structures of synthetic peptides were determined with a JASCO J-815 CD spectrometer (Jasco, Essex, UK). Fifty micromolar peptide samples were, respectively, dissolved in ddH_2O (aqueous environment) and 30 mM sodium dodecyl sulphate (SDS) buffer (membrane-mimetic environment) (Sigma, Dorset, UK). The measurement was completed at 20 °C with the wavelength of 195–250 nm, 1 nm bandwidth, 0.5 nm data pitch and scanning speed of 100 nm/min. The collected results were then analysed with the online analysis tool, K2D3 (http://cbdm-01.zdv.uni-mainz.de/~andrade/k2d3/) [23].

2.5. Trypsin, Chymotrypsin and Tryptase Inhibition Assays

Trypsin and chymotrypsin inhibition assays were performed as previously described [24]. The substrate of trypsin and chymotrypsin were Phe-Pro-Arg-AMC and Succinyl-Ala-Ala-Pro-Phe-NHMec, respectively, and obtained from Bachem, Merseyside, UK. In the tryptase inhibition assay, peptide samples were first dissolved in tryptase assay buffer pH 7.6, containing 0.05 M Tris, 0.15 M NaCl and 0.2% (w/v) PEG 6000 (final volume 210 µL) to obtain different concentrations of 0.5, 1, 2 and 4 mM and then added to the wells of a black 96-well plate containing (Boc-Phe-Ser-Arg-NHMec, obtained from Bachem, Merseyside, UK) (50 µM). Before measurement, tryptase (2.5 µL from 1 mg/mL stock solution, Calbiochem, Nottingham, UK) was added to each well. The rate of hydrolysis of substrate was monitored continuously at 37 °C and 460 nm with a Fluostar Optima plate reader (BMG Labtech GmbH, Offenburg, Germany)

2.6. Minimum Inhibitory Concentration (MIC) and Minimal Bactericidal Concentration (MBC) Assays

The MICs and MBCs of RNF and its analogues were evaluated using seven types of microorganisms as previous described [25]: the Gram-positive bacteria, *Staphylococcus aureus* (*S. aureus*) (NCTC 10788), *Enterococcus faecalis* (*E. faecalis*) (NCTC 12697) and Methicillin-resistant *Staphylococcus aureus* (*MRSA*) (NCTC 12493); the Gram-negative bacteria, *Escherichia coli* (*E. coli*) (NCTC 10418) *Pseudomonas aeruginosa*

(*P. aeruginosa*) (ATTC 27853) and the *Klebsiella pneumoniae (K. pneumoniae)* (ATCC43816); and a single yeast, *Candida albicans (C. albicans)* (NCPF 1467).

2.7. Checkerboard Assays

Peptides at different concentrations (1, 2, 3 and 4 μM) were loaded along the rows of a 96-well plate, while the columns received different concentrations of Gentamicin (0.25, 0.5, 1 and 2 μM) (Sigma, UK). Then, the microorganism cultures (1×10^6 cfu/mL) were inoculated into the plate wells and incubated at 37 °C for 18 h. The optical densities of the cultures were measured using an ELISA plate reader (Biotech Minneapolis, MN, USA). The lowest cumulative fractional inhibitory concentration (ΣFIC) was calculated with the following equation:

$$\Sigma\text{FIC} = \frac{\text{Compound A treated in combination}}{\text{Compound A, test in alone}} + \frac{\text{Compound B treated in combination}}{\text{Compound B, test in alone}} \tag{1}$$

ΣFIC ≤ 0.5 indicates synergy; $0.5 < \Sigma$ FIC ≤ 4 indicates an additive effect; and Σ FIC > 4 represents an antagonistic effect [26–28].

2.8. Haemolysis and Cytotoxicity Assays

The haemolysis assay was performed using horse erythrocytes (2% suspension) (TCS Biosciences Ltd., Buckingham, UK) as reported previously [29]. Serial concentrations of peptide were incubated with blood cell suspension at 37 °C for 2 h. Then, 1% Triton X-100 (Sigma, Dorset, UK) was used to determine total haemolysis of cells as a positive (100%) control of test peptide and added phosphate-buffered saline (PBS) served as the negative control (0%). After this, the mixtures were centrifuged at $1000\times g$ for 5 min and lysis of red cells was determined by measuring of the optical density values of supernatants at 550 nm.

The cytotoxic effects of RNF and its analogues on mammalian cells was examined using a human keratinocyte cell line (HaCaT) (ATCC-PCS-200-011) and a human microvascular endothelial cell line (HMEC-1) (ATCC-CRL-3243). Cells (5×10^3 cells/mL) were seeded into 96-well plates with FBS-containing fresh medium for 24 h. After that, cells were treated with corresponding serum-free culture medium for 12 h. Cells were then treated with RNF in a concentration range from 10^{-9} to 10^{-4} M. After 24 h of treatment, MTT reagent (5 mg/mL) was added to each sample and cells were incubated for 4–6 h at 37 °C. After exposure for 6 h, all liquid was removed and 100 μL of DMSO was added. Finally, the absorbance of samples at 570 nm was determined using an ELISA plate reader (Biotech, Minneapolis, MN, USA).

2.9. Statistical Analysis

Data were assessed for statistical significance using One-Way ANOVA with Prism (Version 6.0; GraphPad Software Inc., San Diego, CA, USA). Results are reported as the mean ± SEM with significance (* $p < 0.05$, ** $p < 0.01$, *** $p < 0.001$ and **** $p < 0.0001$).

3. Results

3.1. Molecular Cloning of the RNF Precursor-Encoding cDNA and Sequence Analysis

Through use of "shotgun" cloning, the RNF precursor cDNA was cloned successfully from the *Pelophylax nigromaculatus* skin secretion cDNA library. As shown in Figure 1, the deduced precursor of RNF consisted of 63 amino acid residues, which included a 22-amino acid residues signal peptide, a 21-mer acidic amino acid spacer and a 17-amino residues mature peptide. The deduced mature peptide was preceded by a typical processing site (-K-R-) and ended in a glycine residue, which is the classical amino donor for C-terminal amidation (Figure 1). The alignment of nucleotide and peptide precursor amino acid sequence exhibited a high degree of similarity between ranacyclin-HB1 and ranacyclin-HB2 (Figure 2). The nucleotide sequence of RNF has been deposited in the GenBank database (accession number MT584032).

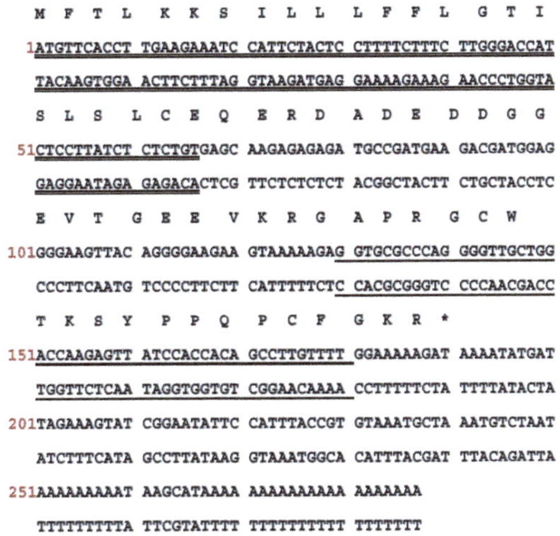

Figure 1. Nucleotide and deduced amino acid sequences of cDNA encoding RNF from *Pelophylax nigromaculatus*. There are 63 amino acid residues within the open-reading frame. The putative signal peptide sequence is double-underlined. The following acidic spacer peptide ends with a typical -K-R- cleavage site. The deduced mature peptide is shown with a single underline and the stop codon is marked with an asterisk.

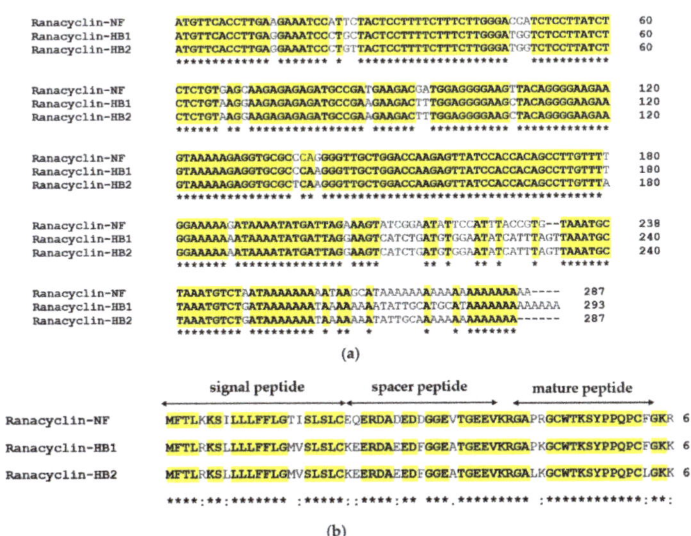

Figure 2. (a) Alignment of precursor cDNA nucleotides of RNF, ranacyclin-HB1 and ranacyclin-HB2; and (b) alignment of precursor cDNA translated peptides of RNF, ranacyclin-HB1 and ranacyclin-HB2. Asterisks indicate identical nucleotides/amino acids (in (a,b)) and colons (:) indicate chemically-conserved amino acids (in (b)).

3.2. Secondary Structure Determinations

After the sequence was unequivocally established, RNF and its two analogues were synthesised by standard solid-phase Fmoc chemistry. Subsequent analyses of MALDI-TOF MS spectra of respective synthesis mixtures showed molecular masses of major components to be consistent with theoretical masses, indicating that all three syntheses of RNF were successfully obtained (Figure 3a–c). The predicted secondary structure of RNF was acquired by use of I-TASSER and visualised with Pymol, and this indicated that this peptide may adopt a random coil structure (Figure 4a) [23,30]. The precise secondary structures of RNF and its analogues were determined by CD spectroscopy in aqueous (ddH$_2$O) and membrane-mimetic environments (SDS buffer), respectively (Figure 4b). The proportion of different secondary structure domains were calculated using the online tool K2D3 webserver (Figure 4c). It appeared that RNF and both of its analogues adopted similar secondary structures in either aqueous or membrane-mimetic environments, which was generally consistent with the predicted model (Figure 4a).

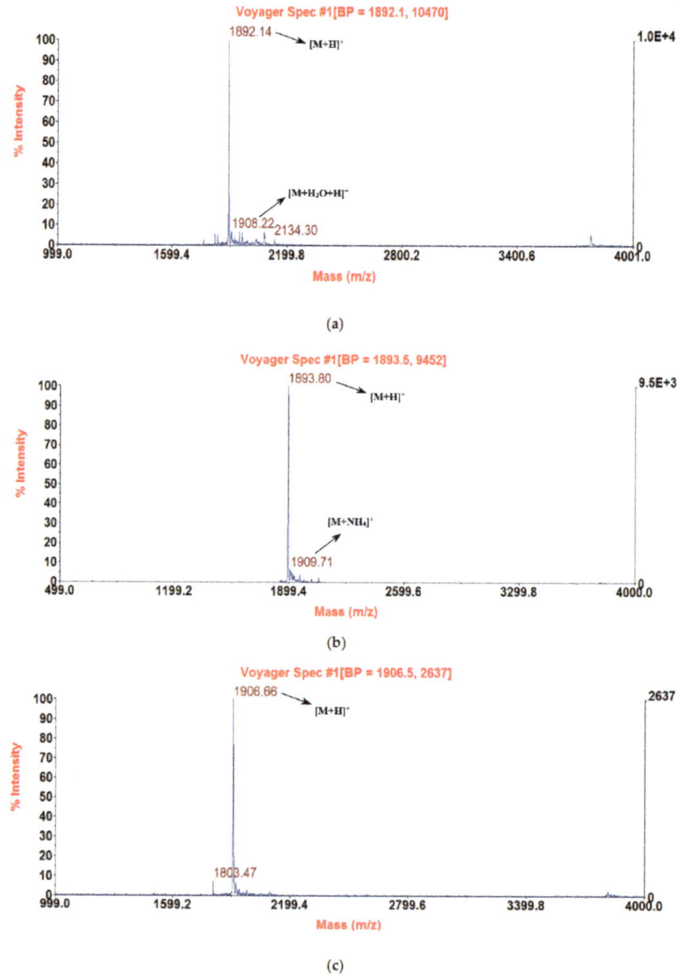

Figure 3. MALDI-TOF MS spectra of the three purified synthetic peptides: (**a**) RNF (1892.69 [M + H]⁺); (**b**) RNF1 (1893.8 [M + H]⁺); and (**c**) RNF3L (1906.66 [M + H]⁺).

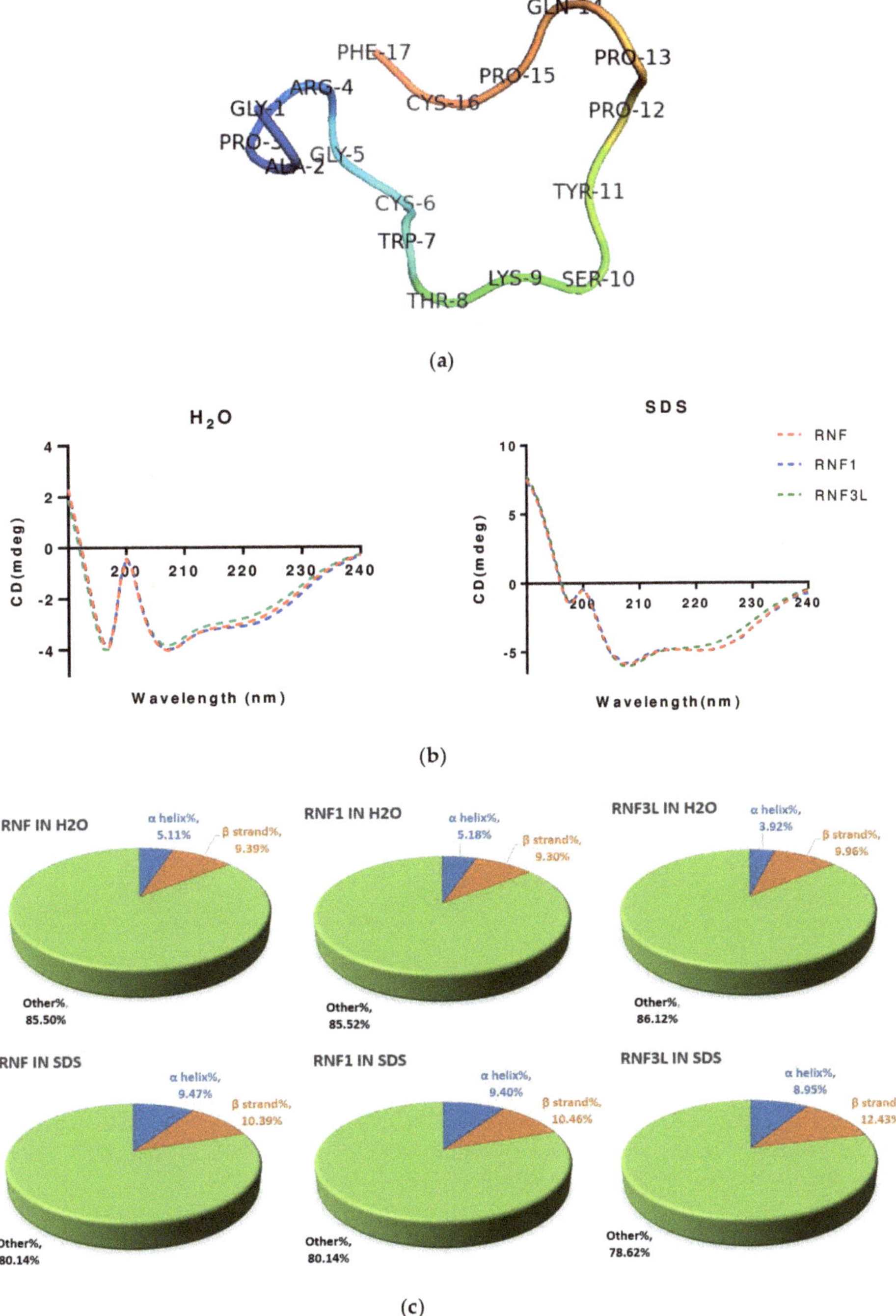

Figure 4. (**a**) Predicted secondary structure of RNF; (**b**) secondary structures of RNF (red lines), RNF1 (blue lines) and RNF3L (green lines) measured in ddH₂O and 30 mM SDS buffers; and (**c**) proportion of different secondary structure domains predicted and calculated by K2D3.

3.3. Protease Inhibitory Activity Assays

RNF and its analogues were subjected to protease inhibitory assays against trypsin, chymotrypsin and tryptase, respectively. RNF exhibited obvious trypsin inhibitory activity with a Ki value of 0.447 μM. The trypsin inhibitory activity of RNF1 was lower with a Ki of 1.3 μM. RNF3L showed the strongest trypsin inhibitory activity among these three with a Ki of 0.201 μM. They showed similar tryptase inhibitory activity but no chymotrypsin inhibitory activity (Table 1). Compared with other ranacyclin-type peptides, although peptides in this work shared high sequence identity with merely two or three residues mutated, their trypsin inhibitory activity was surprisingly different.

Table 1. Inhibitory activity of RNF, RNF1, RNF3L and some ranacyclins against trypsin, chymotrypsin and tryptase.

Peptide	Peptide Sequence	Ki(μM) of Trypsin	Ki(μM) of Chymotrypsin	Ki(μM) of Tryptase
RNF	GAPRGCWTKSYPPQPCF-NH$_2$	0.447	N.I.	6.774
RNF1	GAPRGCWTKSYPPQPCF	1.300	N.I.	9.059
RNF3L	GALRGCWTKSYPPQPCF-NH$_2$	0.201	N.I.	12.5
Ranacyclin-T [10]	GALRGCWTKSYPPKPCK	0.116	-	-
HJTI [31]	GAPKGCWTKSYPPQPCS-NH$_2$	0.388	-	-
OSTI [24]	AALKGCWTKSIPPKPCF-NH$_2$	0.0003	N.I.	2.5
PPF-BBI [32]	ALRGCWTKSIPPKPCP-NH$_2$	0.17	N.I.	30.73

N.I., there was no inhibitory activity; -, there was no mention in previous studies.

3.4. Antimicrobial Activity

Antimicrobial activity of RNF and its analogues against Gram-positive bacteria, Gram-negative bacteria and fungi were determined by using a doubling-dilution method. RNF showed a weak bacteriostatic activity toward *S. aureus* at 512 μM, but other peptides did not exhibit obvious antimicrobial activity against any tested microorganisms even at the highest concentration of 512 μM employed (Table 2).

Table 2. Antimicrobial activity of RNF and its analogues.

	MICs/MBCs (μM)						
Peptide	*S. aureus*	*MRSA*	*E. faecalis*	*E. coli*	*P. aeruginosa*	*K. pneumoniae*	*C. albicans*
RNF	512/>512	>512/>512	>512/>512	>512/>512	>512/>512	>512/>512	>512/>512
RNF1	>512/>512	>512/>512	>512/>512	>512/>512	>512/>512	>512/>512	>512/>512
RNF3L	>512/>512	>512/>512	>512/>512	>512/>512	>512/>512	>512/>512	>512/>512

3.5. Additive Antimicrobial Effects with Gentamicin

Combinations of an antibiotic (Gentamicin) with either RNF or one of its analogues, improved the activity of the peptides against MRSA (Figure 5). As shown in the following figure, the MIC of Gentamicin against MRSA was 2 μM. In the presence of RNF and its analogues, the MIC of Gentamicin against *MRSA* was reduced to 1 μM. Since RNF and its analogues showed no obvious inhibitory activity against *MRSA* at 512 μM, the lowest cumulative fractional inhibitory concentration (ΣFIC) would be between 0.5 and 1, indicative of an additive effect between peptide and Gentamicin. The best combination of peptide with Gentamicin was 1 μM with 1 μM, respectively.

3.6. Haemolysis and Cytotoxicity

Released haemoglobin served as a signal of haemolytic activity of peptide. The absorbance of red blood cell lysis was evaluated at a wavelength of 570 nm. RNF and analogues were found to have a weak haemolytic activity and induced less than 5% haemolysis at a concentration of 512 μM (Figure 5d). Moreover, the peptides showed negligible cytotoxicity toward human normal cell lines even at the highest concentrations tested of 100 μM (Figure 6).

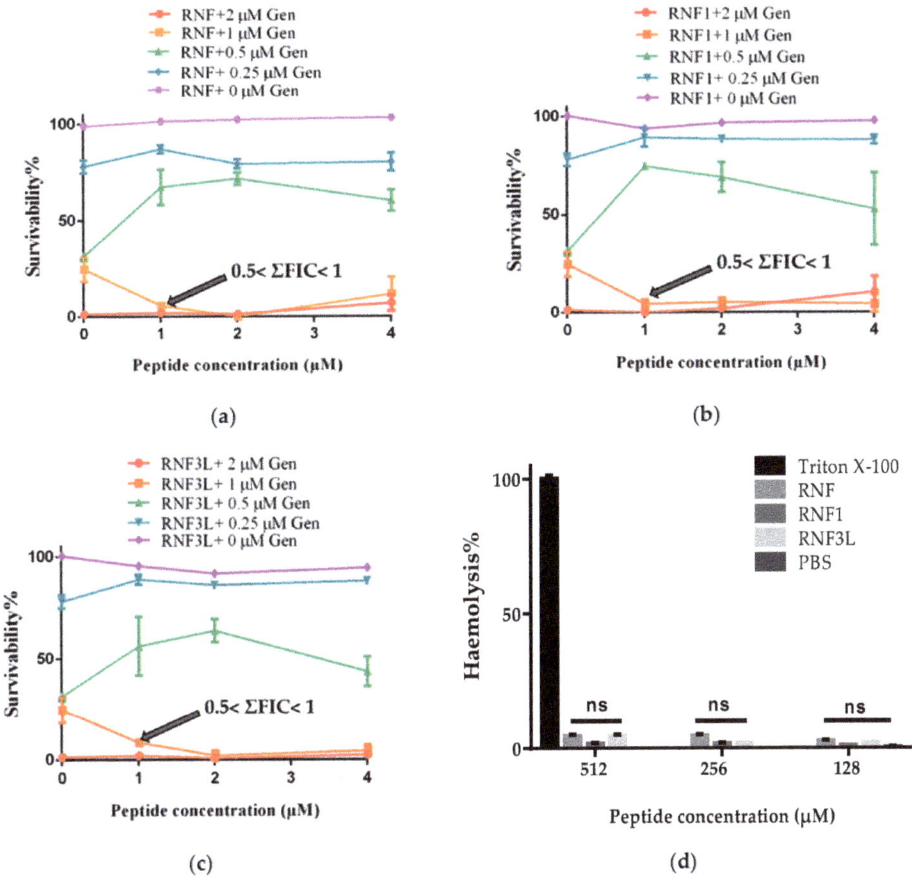

Figure 5. The additive effect against the growth of MRSA between Gentamicin (Gen) and: (**a**) RNF; (**b**) RNF1; and (**c**) RNF3L. (**d**) Haemolysis of RNF, RNF1 and RNF3L at 128, 256 and 512 µM. PBS and 1% Triton X-100 were set as negative and positive controls, respectively.

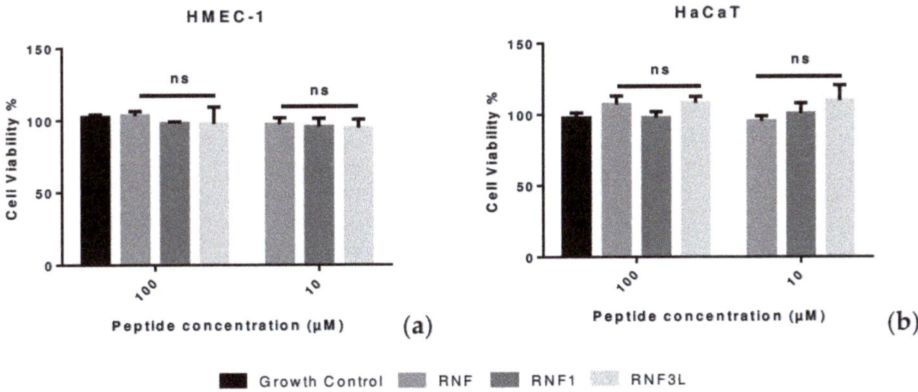

Figure 6. The cytotoxicity of (RNF) RNF1 and RNF3L: on HMEC-1 cells (**a**); and on HaCaT cells (**b**).

4. Discussion

In previous studies, extensive studies have been made to reveal the role played by each residue within the TIL in overall enzyme inhibitory activity. However, factors outside this conserved disulphide loop, which may affect trypsin inhibition activity, have rarely been studied. In 2012, Wang et al. discovered the peptide, HJTI, and designed one molecular variant to study the role that cationicity played in trypsin inhibitory potency. They believed that structural features, especially for amino acid residues lying outside the TIL, were crucial in modulating the effectiveness of trypsin inhibitory activity [31]. In addition, according to a previous report, ranacyclin-T characterised from *Rana esculenta*, showed a stronger trypsin inhibitory activity than RNF with a Ki of 116 nM [10]. To construct rational explanations for the different trypsin inhibitory activities between ranacyclin-T and RNF and to further study structure and trypsin inhibitory activity relationships, we compared some structural parameters of these two peptides (Table 3). Compared with RNF, ranacyclin-T contained a larger net positive charge and a different residue (Lys) at the P5′ position—a position thought to play important role in the trypsin inhibitory activity of naturally-occurring BBIs [4]. Analysis of predicted secondary structures also revealed that ranacyclin-T may adopt a beta-sheet structure which was different for RNF3L (Figure 7). Taken together, the different trypsin inhibitory activities of RNF and ranacyclin-T could be generally explained by three points: (1) different secondary structures; (2) different cationicities; and (3) different residues at the P5′ position. In a previous study, amino acid scanning at the P5′ position within the BBIs inhibitory loop, revealed that the residue Gln (Q) was the optimal residue at the P5′ position with a relatively higher inhibitory activity in both bovine beta-trypsin and human cationic trypsin inhibitory studies [4]. Therefore, residues at the P5′ position might not be the main reason for induction of different activities. On consideration of the effect of cationicity, which may play a minor role in trypsin inhibitory potency [31], it was assumed that secondary structure may contribute to different trypsin inhibitory activities. The different secondary structures of these two peptides were possibly induced by the different residues at the third position since Pro usually inhibits backbones to conform to alpha-helix or beta-sheet structures [33]. Therefore, it was decided here to focus on the possible role of features outside the TIL in trypsin inhibitory activity. To this end, the structure of ranacyclin-T was used as a template to generate RNF3L, which was designed by replacing the third residue of Pro with Leu, aiming to study potential relationships of structure–activity. Moreover, for some known AMPs, C-terminal amidation is considered to play a significant role in structural stability and antimicrobial activity. In addition, according to Li et al.'s work in 2007, the C-terminal naked phenylalanine of ORB may help promote antimicrobial efficacy due to its strong hydrophobicity, which may facilitate the aggregation of peptide to form a channel-like structure on cell membranes [21]. Therefore, the free C-terminal analogue, RNF1, was synthesised to study the role of C-terminal amidation in trypsin inhibitory capacity and antimicrobial activity.

Table 3. Some parameters of RNF and some reported ranacyclins and their trypsin inhibition Ki values. Amino acid residues at P5′ are shaded.

Peptide	Peptide Sequence	Net Charge	Trypsin Inhibition Ki (nM)
RNF	GAPRGCWTKSYPPQPCF-NH$_2$	3	447
Ranacyclin-T [10]	GALRGCWTKSYPPKPCK-NH$_2$	5	116
HJTI [31]	GAPKGCWTKSYPPQPCS-NH$_2$	3	388
OSTI [24]	AALKGCWTKSIPPKPCF-NH$_2$	4	0.3
PPF-BBI [32]	ALRGCWTKSIPPKPCP-NH$_2$	4	170

In the trypsin inhibition assay, RNF showed obvious inhibitory activity with a Ki of 447 nM. Compared with RNF, RNF1 exhibited relatively lower inhibitory potency with a Ki of 1.3 μM. Some researchers believe that C-terminal amidation plays a fundamental role in peptide stability since the terminal amidation could generate a mimic of native proteins [1]. Therefore, the lack of C-terminal amidation might make RNF1 more flexible in structure, which might be unfavourable for peptide to combine with reactive sites on trypsin. RNF3L showed the strongest trypsin inhibitory activity in assay with a Ki value of 201 nM. The increased trypsin inhibitory activity of RNF3L may be due to two reasons: (1) the residue Leu might be more suitable than Pro at this position; and (2) the formation of a beta-sheet leads to the improvement of inhibitory activity. According to the results of CD spectroscopy, RNF and its analogues adopted similar secondary structures both in aqueous and membrane-mimetic environments. Therefore, it is unlikely that secondary structure contributes to the improvement of trypsin inhibitory activity shown by RNF3L. To shed light on exactly how RNF3L shows stronger trypsin inhibitory activity requires the design and performance of additional experiments in the future. In addition, more peptides need to be identified or designed to make clear the exact roles of residues outside the TIL on the exhibition of trypsin inhibitory activity. Taken together, both the C-terminal de-amidation effect on increased trypsin-inhibition activity and the substitution of specific residues in the peptide to this end, provided several lines of evidence to the hypothesis that structural features outside the disulphide loop may play a more fundamental role in dictating inhibitory potency. In the chymotrypsin inhibitory assay, none of the tested peptides showed inhibitory activity. In fact, according to the specific residues at cleavage site P1, the serine protease could be categorised into trypsin, chymotrypsin and elastase analogues. For trypsin-like proteases, the S1 pocket is narrow and anionic. However, the S1 pocket of chymotrypsin-like proteases is broad and hydrophobic [34]. Therefore, chymotrypsin inhibitors or activators generally contain specific residues, such as Trp, Phe or Leu. In this report, the residue at the P1 position of RNF and its analogues was a Lys, which dictated they cannot inhibit chymotrypsin. Modification work in this study revealed the importance of residues and structural features outside TIL for the trypsin inhibitory activity of the ranacyclins peptide. A good understanding of the structure and trypsin inhibitory activity relationship would be favourable for the designs of improved peptides, which are promising candidates in the treatment of health problems, such as metabolic syndrome [35].

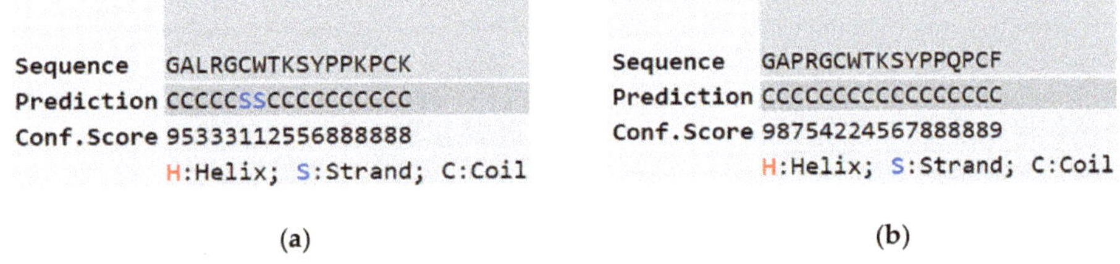

(a)   (b)

Figure 7. Predicted secondary structures of ranacyclin-T (**a**) and RNF (**b**), obtained from I-TASSER (https://zhanglab.ccmb.med.umich.edu/I-TASSER/) [30].

Tryptases have been found to play important roles in some diseases, such as atherosclerosis, mastocytosis and acute myeloid leukemia. Tryptases in mast cells have also been shown to be closely related to the symptoms of allergic asthma and inflammation [10]. Therefore, the emergence of a novel tryptase inhibitor would be an asset in the treatment of some clinical diseases. In this study, RNF was found to have obvious tryptase inhibitory activity, which may help to solve some difficulties in the clinic. Some publications believe that protease inhibitors cannot work effectively when it comes to human proteases [36]. However, RNF and its designed analogues, as well as some other reported BBI peptides such as OSTI and PPF-BBI, show obvious tryptase inhibitory activity, which was different to previous reports [24,32]. It could be that the larger protease inhibitors might be too large to approach the reactive sites on tryptase and if this should be the case, then some smaller protease inhibitors, such as peptides of the BBI family, may have a great advantage in accessing these active sites.

Ranacyclins were defined as a novel antimicrobial peptide family with potent antibacterial and anti-fungi activities; unlike most cationic AMPs, ranacyclins tend to adopt a pre-dominant random coil structure and can bind and insert into the membrane primarily through the interaction with the hydrophobic core of the bacterial membrane rather than electrostatic interaction [13]. Therefore, ranacyclins firstly approach to the membrane surface through hydrophobic interaction and then, at sites where peptide concentrations reach a threshold, insert into the hydrophobic core of the membrane bilayer and form a channel-like structure leaking the cells [13,37]. However, RNF in this work did not exhibit obvious bacteriostatic or bactericidal effects as other reported ranacyclins. Except for the weak bacteriostatic activity of RNF against *S. aureus*, RNF and its variants showed no obvious antimicrobial activity. These "abnormal" results, nevertheless, were not sufficient to prove that RNF was a "fake" antimicrobial peptide or not all ranacyclins exhibit antimicrobial properties. Bacteria used in this reported were different from those in previous papers; RNF without antimicrobial efficacy in this study does not mean it cannot kill other untested strains. To comprehensively evaluate the potential antimicrobial efficacy of RNF or ranacyclins, current studies were inadequate and more bacterial strains need to be employed. In this study, RNF adopted a mainly random coil structure and had a low cationicity. Ranacyclin-T reported previously adopted a beta-sheet structure, displayed a higher net positive charge and had relatively stronger antimicrobial activity with respect to RNF, which may prove that different structural proportions and cationicity may have contributed to the negligible antimicrobial activity of RNF [13]. For RNF1, although C-terminal de-amidation exposed a naked phenylalanine residue, it also came with a reduction in net charge, which may explain why RNF1 could not inhibit the growth of test microorganisms in this study. Although RNF with weak bacteriostatic activities were hard to be applied in the pharmaceutical fields, their negligible cytotoxicity and obvious trypsin inhibitory activity allow them to be applied in the food industry where the requirements for microbiota control are not as strict as in clinical helping to control pressing obesity problems [35,38]. Moreover, using the advantages of multiple functions, RNF could serve as a temple to design some ameliorated derivatives [21,38].

Gentamicin is a common antibiotic which is effective against a wide range of microorganisms, including Gram-positive bacteria and Gram-negative bacteria. Due to its excellent antimicrobial efficacy and broad-spectrum action, Gentamicin is widely-employed in clinical treatment [39]. However, as a typical aminoglycoside antibiotic, Gentamicin is considered as an "obligatory nephrotoxin" and even small doses could lead to the production of nephrotoxicity in humans and animals [40]. Therefore, the nephrotoxicity from aminoglycosides was considered as an inevitable barrier for patients. On the one hand, in a previous study, BBIs were found to mitigate Gentamicin-induced nephrotoxicity while not decreasing the therapeutic effects of Gentamicin [41]. On the other hand, some peptides in the BBI family were reported to show synergistic antimicrobial effects with some clinical antibiotics, such as rifampicin and erythromycin [13]. Therefore, in this study, Gentamicin was used to investigate whether RNF could help improve its therapeutic effect. The antimicrobial ability of Gentamicin was improved in the presence of RNF with a lower MIC against MRSA. Moreover, RNF and its analogues showed no obvious haemolysis (inducing no more than 5% haemolysis at 512 μM) and negligible cytotoxicity toward human normal cell lines, HMEC-1 and HaCaT. Considering the possible decreasing of nephrotoxicity, low toxic effects per se and improved antimicrobial efficacy, the application of RNF was promising and deserving of further critical study.

5. Conclusions

A novel BBI peptide, RNF, was characterised from the skin secretion of *Pelophylax nigromaculatus*. The different trypsin inhibitory activities of RNF1 and RNF3L revealed residues or structures outside of the trypsin inhibition loop per se, fundamentally contribute to their trypsin inhibitory activity. To prove this hypothesis, more experiments remain to be designed and completed. Although RNF and its analogues showed no obvious bactericidal activity in antimicrobial assays, they could help to improve the therapeutic efficacy of Gentamicin or other conventional antibiotics, at low concentrations. Moreover, these three peptides showed negligible haemolysis and cytotoxicity toward normal cells. Therefore, the further analysis and perhaps development of RNF may be of value in addressing or solving some pressing health problems.

Author Contributions: Conceptualisation, L.W., M.Z. and T.C.; Data curation, Y.J. and T.W.; Formal analysis, T.W. and X.C.; Investigation, T.W., Y.J. and Y.Z.; Methodology, X.X. and C.M.; Project administration, M.Z.; Resources, M.Z. and Y.Z.; Supervision, X.C., L.W., M.Z. and T.C.; and Validation, X.C. and L.W. Writing—original draft, T.W., X.C. and L.W.; Writing—review & editing, Y.Z., C.S. and M.Z. All authors have read and agreed to the published version of the manuscript.

Funding: This research received no external funding.

Conflicts of Interest: The authors declare no conflict of interest.

References

1. Hedstrom, L. Serine Protease Mechanism and Specificity. *Chem. Rev.* **2002**, *102*, 4501–4524. [CrossRef] [PubMed]

2. Bachovchin, D.; Cravatt, B. The pharmacological landscape and therapeutic potential of serine hydrolases. *Nat. Rev. Drug Discov.* **2012**, *11*, 52–68. [CrossRef] [PubMed]

3. Drag, M.; Salvesen, G. Emerging principles in protease-based drug discovery. *Nat. Rev. Drug Discov.* **2010**, *9*, 690–701. [CrossRef] [PubMed]

4. Li, C.; de Veer, S.; White, A.; Chen, X.; Harris, J.; Swedberg, J.; Craik, D. Amino Acid Scanning at P5′ within the Bowman–Birk Inhibitory Loop Reveals Specificity Trends for Diverse Serine Proteases. *J. Med. Chem.* **2019**, *62*, 3696–3706. [CrossRef]

5. Qi, R.; Song, Z.; Chi, C. Structural Features and Molecular Evolution of Bowman-Birk Protease Inhibitors and Their Potential Application. *Acta Biochim. Biophys. Sin.* **2005**, *37*, 283–292. [CrossRef]

6. BIRK, Y. The Bowman-Birk inhibitor. Trypsin- and chymotrypsin-inhibitor from soybeans. *Int. J. Pept. Protein Res.* **1985**, *25*, 113–131. [CrossRef]

7. McBride, J.; Leatherbrrow, R. Synthetic Peptide Mimics of the Bowman-Birk Inhibitor Protein. *Curr. Med. Chem.* **2001**, *8*, 909–917. [CrossRef]

8. Lin, Y.; Hang, H.; Chen, T.; Zhou, M.; Wang, L.; Shaw, C. pLR-HL: A Novel Amphibian Bowman-Birk-type Trypsin Inhibitor from the Skin Secretion of the Broad-folded Frog, *Hylarana latouchii*. *Chem. Biol. Drug Des.* **2016**, *87*, 91–100. [CrossRef]

9. Li, J.; Xu, X.; Xu, C.; Zhou, W.; Zhang, K.; Yu, H.; Zhang, Y.; Zheng, Y.; Rees, H.; Lai, R.; et al. Anti-infection Peptidomics of Amphibian Skin. *Mol. Cell Proteom.* **2007**, *6*, 882–894. [CrossRef]

10. Rothemund, S.; Sönnichsen, F.; Polte, T. Therapeutic Potential of the Peptide Leucine Arginine as a New Nonplant Bowman–Birk-Like Serine Protease Inhibitor. *J. Med. Chem.* **2013**, *56*, 6732–6744. [CrossRef]

11. McBride, J.; Watson, E.; Brauer, A.; Jaulent, A.; Leatherbarrow, R. Peptide mimics of the Bowman-Birk inhibitor reactive site loop. *Biopolymers* **2002**, *66*, 79–92. [CrossRef] [PubMed]

12. Safavi, F.; Rostami, A. Role of serine proteases in inflammation: Bowman–Birk protease inhibitor (BBI) as a potential therapy for autoimmune diseases. *Exp. Mol. Pathol.* **2012**, *93*, 428–433. [CrossRef]

13. Mangoni, M.; Papo, N.; Mignogna, G.; Andreu, D.; Shai, Y.; Barra, D.; Simmaco, M. Ranacyclins, a New Family of Short Cyclic Antimicrobial Peptides: Biological Function, Mode of Action, and Parameters Involved in Target Specificity. *Biochemistry* **2003**, *42*, 14023–14035. [CrossRef]

14. Xu, X.; Lai, R. The chemistry and biological activities of peptides from amphibian skin secretions. *Chem. Rev.* **2015**, *115*, 1760–1846. [CrossRef] [PubMed]

15. Graham, C.; Irvine, A.E.; McClean, S.; Richter, S.C.; Flatt, P.R.; Shaw, C. Peptide Tyrosine Arginine, a potent immunomodulatory peptide isolated and structurally characterized from the skin secretions of the dusky gopher frog, *Rana sevosa*. *Peptides* **2005**, *26*, 737–743. [CrossRef] [PubMed]

16. Salmon, A.L.; Cross, L.J.; Irvine, A.E.; Lappin, T.R.; Dathe, M.; Krause, G.; Canning, P.; Thim, L.; Beyermann, M.; Rothemund, S.; et al. Peptide leucine arginine, a potent immunomodulatory peptide isolated and structurally characterized from the skin of the Northern Leopard frog, *Rana pipiens*. *J. Biol. Chem.* **2001**, *276*, 10145–10152. [CrossRef]

17. Hao, X.; Tang, X.; Luo, L.; Wang, Y.; Lai, R.; Lu, Q. A novel ranacyclin-like peptide with anti-platelet activity identified from skin secretions of the frog *Amolops loloensis*. *Gene* **2016**, *576*, 171–175. [CrossRef]

18. Wang, X.; Ren, S.; Guo, C.; Zhang, W.; Zhang, X.; Zhang, B.; Li, S.; Ren, J.; Hu, Y.; Wang, H. Identification and functional analyses of novel antioxidant peptides and antimicrobial peptides from skin secretions of four East Asian frog species. *Acta Biochim. Biophys. Sin.* **2017**, *49*, 550–559. [CrossRef]

19. Yang, H.; Wang, X.; Liu, X.; Wu, J.; Liu, C.; Gong, W.; Zhao, Z.; Hong, J.; Lin, D.; Wang, Y.; et al. Antioxidant peptidomics reveals novel skin antioxidant system. *Mol. Cell Proteom.* **2009**, *8*, 571–583. [CrossRef]

20. Demori, I.; Rashed, Z.E.; Corradino, V.; Catalano, A.; Rovegno, L.; Queirolo, L.; Salvidio, S.; Biggi, E.; Zanotti-Russo, M.; Canesi, L.; et al. Peptides for Skin Protection and Healing in Amphibians. *Molecules* **2019**, *24*, 347. [CrossRef]

21. Li, J.; Zhang, C.; Xu, X.; Wang, J.; Yu, H.; Lai, R.; Gong, W. Trypsin inhibitory loop is an excellent lead structure to design serine protease inhibitors and antimicrobial peptides. *FASEB J.* **2007**, *21*, 2466–2473. [CrossRef]

22. Yuan, Y.; Zai, Y.; Xi, X.; Ma, C.; Wang, L.; Zhou, M.; Shaw, C.; Chen, T. A novel membrane-disruptive antimicrobial peptide from frog skin secretion against cystic fibrosis isolates and evaluation of anti-MRSA effect using *Galleria mellonella* model. *Biochim. Biophys. Acta Gen. Subj.* **2019**, *1863*, 849–856. [CrossRef] [PubMed]

23. Louis-Jeune, C.; Andrade-Navarro, M.A.; Perez-Iratxeta, C. Prediction of protein secondary structure from circular dichroism using theoretically derived spectra. *Proteins* **2012**, *80*, 374–381. [CrossRef]

24. Wu, Y.; Long, Q.; Xu, Y.; Guo, S.; Chen, T.; Wang, L.; Zhou, M.; Zhang, Y.; Shaw, C.; Walker, B. A structural and functional analogue of a Bowman–Birk-type protease inhibitor from Odorrana schmackeri. *Biosci. Rep.* **2017**, *37*. [CrossRef] [PubMed]

25. Li, M.; Xi, X.; Ma, C.; Chen, X.; Zhou, M.; Burrows, J.F.; Chen, T.; Wang, L. A Novel Dermaseptin Isolated from the Skin Secretion of Phyllomedusa tarsius and Its Cationicity-Enhanced Analogue Exhibiting Effective Antimicrobial and Anti-Proliferative Activities. *Biomolecules* **2019**, *9*, 628. [CrossRef] [PubMed]

26. Xiang, J.; Zhou, M.; Wu, Y.; Chen, T.; Shaw, C.; Wang, L. The synergistic antimicrobial effects of novel bombinin and bombinin H peptides from the skin secretion of *Bombina orientalis*. *Biosci. Rep.* **2017**, *37*, BSR20170967. [CrossRef]

27. Odds, F.C. Synergy, antagonism, and what the chequerboard puts between them. *J. Antimicrob. Chemother.* **2003**, *52*, 1. [CrossRef]

28. Hall, M.J.; Middleton, R.F.; Westmacott, D. The fractional inhibitory concentration (FIC) index as a measure of synergy. *J. Antimicrob. Chemother.* **1983**, *11*, 427–433. [CrossRef]

29. Jiang, Y.; Wu, Y.; Wang, T.; Chen, X.; Zhou, M.; Ma, C.; Xi, X.; Zhang, Y.; Chen, T.; Shaw, C.; et al. Brevinin-1GHd: A novel *Hylarana guentheri* skin secretion-derived Brevinin-1 type peptide with antimicrobial and anticancer therapeutic potential. *Biosci. Rep.* **2020**, *40*, BSR20200019. [CrossRef]

30. Roy, A.; Kucukural, A.; Zhang, Y. I-TASSER: A unified platform for automated protein structure and function prediction. *Nat. Protoc.* **2010**, *5*, 725–738. [CrossRef]

31. Wang, M.; Wang, L.; Chen, T.; Walker, B.; Zhou, M.; Sui, D.; Conlon, J.; Shaw, C. Identification and molecular cloning of a novel amphibian Bowman Birk-type trypsin inhibitor from the skin of the Hejiang Odorous Frog; *Odorrana hejiangensis*. *Peptides* **2012**, *33*, 245–250. [CrossRef] [PubMed]

32. Miao, Y.; Chen, G.; Xi, X.; Ma, C.; Wang, L.; Burrows, J.; Duan, J.; Zhou, M.; Chen, T. Discovery and Rational Design of a Novel Bowman-Birk Related Protease Inhibitor. *Biomolecules* **2019**, *9*, 280. [CrossRef] [PubMed]

33. Morgan, A.; Rubenstein, E. Proline: The Distribution, Frequency, Positioning, and Common Functional Roles of Proline and Polyproline Sequences in the Human Proteome. *PLoS ONE* **2013**, *8*, e53785. [CrossRef] [PubMed]

34. Groß, A.; Hashimoto, C.; Sticht, H.; Eichler, J. Synthetic Peptides as Protein Mimics. *Front. Bioeng. Biotechnol.* **2016**, *3*, 211. [CrossRef] [PubMed]

35. Cristina Oliveira de Lima, V.; Piuvezam, G.; Leal Lima Maciel, B.; Heloneida de Araújo Morais, A. Trypsin inhibitors: Promising candidate satietogenic proteins as complementary treatment for obesity and metabolic disorders? *J. Enzym. Inhib. Med. Chem.* **2019**, *34*, 405–419. [CrossRef]

36. Ware, J.H.; Wan, X.S.; Rubin, H.; Schechter, N.M.; Kennedy, A.R. Soybean Bowman-Birk protease inhibitor is a highly effective inhibitor of human mast cell chymase. *Arch. Biochem. Biophys.* **1997**, *344*, 133–138. [CrossRef]

37. Conlon, J.M.; Kolodziejek, J.; Nowotny, N. Antimicrobial peptides from the skins of North American frogs. *Biochim. Biophys. Acta* **2009**, *1788*, 1556–1563. [CrossRef]

38. Cancelarich, N.L.; Wilke, N.; Fanani, M.A.L.; Moreira, D.C.; Pérez, L.O.; Alves Barbosa, E.; Plácido, A.; Socodato, R.; Portugal, C.C.; Relvas, J.B.; et al. Somuncurins: Bioactive Peptides from the Skin of the Endangered Endemic Patagonian Frog Pleurodema somuncurense. *J. Nat. Prod.* **2020**, *83*, 972–984. [CrossRef]

39. Smith, A.; Smith, D. Gentamicin: Adenine Mononucleotide Transferase: Partial Purification, Characterization, and Use in the Clinical Quantitation of Gentamicin. *J. Infect. Dis.* **1974**, *129*, 391–401. [CrossRef]
40. Hariprasad, G.; Kumar, M.; Rani, K.; Kaur, P.; Srinivasan, A. Aminoglycoside induced nephrotoxicity: Molecular modeling studies of calreticulin-gentamicin complex. *J. Mol. Model.* **2012**, *18*, 2645–2652. [CrossRef]
41. Smetana, S.; Khalef, S.; Bar-Khayim, Y.; Birk, Y. Antimicrobial Gentamicin Activity in the Presence of Exogenous Protease Inhibitor (Bowman-Birk Inhibitor) in Gentamicin-Induced Nephrotoxicity in Rats. *Nephron* **1992**, *61*, 68–72. [CrossRef] [PubMed]

MDPI
St. Alban-Anlage 66
4052 Basel
Switzerland
Tel. +41 61 683 77 34
Fax +41 61 302 89 18
www.mdpi.com

Biology Editorial Office
E-mail: biology@mdpi.com
www.mdpi.com/journal/biology